Artificial Intelligence Odyssey: Exploring the Depths of Artificial Intelligence, from Genesis to Generative AI

Robert C. Byrum

.

Introduction

The intersection of innovation and imagination has given rise to a groundbreaking force that transcends the realms of human creativity—the captivating potential of Artificial Intelligence (AI). The possibilities are endless, with AI capable of painting portraits, composing symphonies, and drafting stories that move the soul. This is the frontier we currently stand upon, where AI breathes life into the canvas of creativity and blurs the lines between human ingenuity and machine intelligence.

Join me on an expedition that explores the limitless potential of AI, where algorithms transform into architects of tomorrow's realities. We'll delve into generative AI, an enchanting facet of machine intelligence, and explore the inner workings of neural networks, generative adversarial networks (GANs), variational autoencoders (VAEs), and reinforcement learning. These technological marvels are the engines propelling us into uncharted territories of creative expression.

Our exploration doesn't stop at technical marvels; we'll witness AI's brushstrokes across art, design, and music. From symphonies orchestrated by algorithms to visual arts that defy conventional aesthetics, AI's contributions redefine the essence of creativity.

We'll also journey through AI's applications in industry and innovation, witnessing how it catalyzes disruptive change across sectors, redefining efficiency, and sculpting new paradigms. We pause to ponder the ethical dilemmas and societal implications, considering questions of responsibility and governance in an era where machines craft beauty and utility.

Flow on this expedition as pioneers navigating uncharted terrain. We'll uncover the frontiers of human-AI collaboration, envisioning futures where human creativity harmonizes with the precision of AI intelligence.

The blank canvas, untold melody, and unwritten story of tomorrow lie within the boundless realm of AI's potential. Here, we can sculpt a future where creativity knows no bounds, crafted collaboratively with the synergy between human ingenuity and machine intelligence. Embark with us on this exhilarating quest, as we discover the limitless potential of AI, unshackling its power to craft a tomorrow that defies imagination.

Table of Content

Chapter 7: Generative AI and AI Art

CHAPTER 1

- Genesis of AI

Artificial intelligence's applications and reach don't require a formal introduction. Artificial intelligence is now a reality that permeates every aspect of our daily life, not simply a trendy term. Artificial intelligence (AI) is transforming business sectors like never before as businesses use it to construct clever robots for a variety of applications. This essay on the types of artificial intelligence will teach you about the different stages and classifications of artificial intelligence.

Artificial intelligence: what is it?

The manner of making smart machines from big quantities of statistics is referred to as synthetic intelligence. Systems mimic human behavior by learning from prior mistakes and experiences. It increases human effort's efficacy, speed, and precision. AI builds computers with the ability to make decisions by itself

using sophisticated algorithms and techniques.
Today, artificial intelligence is utilized in practically every business sector:

What influence will artificial intelligence (AI) have on how power is distributed? AI is a general-purpose technology (GPT) having a wide range of uses in both the military and civilian sectors. As a result, a wide range of entities, including nations and businesses, are actively pushing AI innovation and development. While some applications of AI may strengthen current powers, the general-purpose nature of AI will limit first-mover advantages in most AI application areas, especially with regard to balance of power, according to the history of economic and military power. Furthermore, successful AI applications in the military sector would necessitate an organizational transformation that established military forces have found difficult in the past.

History of AI

The purpose of synthetic intelligence (AI), an exceptionally new area that has been around for 60 years, is to imitate human cognitive capacities via a lot of sciences, ideas, and techniques (such as computer science, statistical analysis, probability, and mathematical logic). Started during the Second World War, its advancements are closely related to computing and have allowed computers to accomplish a growing number of intricate jobs that were previously limited to human performance. Nonetheless, this automation continues to be very exceptional from human intelligence withinside the strict sense, that's why a few teachers disagree with the term. Their ultimate goal—a "strong" AI that can solve a wide range of specialized problems in an entirely autonomous manner) is in no way similar to the accomplishments of today ("weak" or "moderate" AIs, incredibly effective in their training field).

For the "strong" AI that has simplest but been visible in technology fiction as a way to version the whole world, fundamental studies improvements instead of simplest overall performance profits would be necessary.

But from 2010, the industry has seen a fresh rise, mostly because of the significant advancements in computer power and the availability of vast amounts of data.

Reaffirmed promises and occasionally imagined worries obstruct a dispassionate comprehension of the phenomena. A brief review of the discipline's past can assist frame the present discussions.

The period between 1940 and 1960 was a time of significant technological developments. The Second World War acted as a catalyst for many innovations, and during this time, there was also a growing interest in understanding how machines and organic beings worked together. Norbert Wiener, a pioneer in cybernetics, aimed to unify mathematical theory, electronics, and automation as "a whole theory of control and communication, both in animals and machines". In 1943, Warren McCulloch and Walter Pitts developed the first mathematical and computer model of the biological neuron (formal neuron).

In the early 1950s, two prominent researchers, John Von Neumann and Alan Turing, played a significant role in the development of the technology behind artificial intelligence (AI). They transitioned from computers that used 19th-century decimal logic (which dealt with values from 0 to 9) to machines that used binary logic (which relied on Boolean algebra and dealt with chains of 0 or 1). Additionally, they formalized the architecture of contemporary computers and demonstrated that it was a universal machine capable of executing what is programmed.

In 1950, Turing raised the question of whether machines could possess intelligence in his famous article "Computing Machinery and Intelligence. He described a "game of imitation" in which a human should be able to distinguish in a teletype dialogue whether they are talking to a man or a machine. Although this article remains controversial, it is often cited as the origin of the debate surrounding the boundary between humans and machines.

Overall, this period saw significant advancements in technology and laid the foundation for the development of AI. The work of researchers such as Wiener, McCulloch, Pitts, Von Neumann, and Turing helped shape the field and set the stage for future progress.

Artificial Intelligence (AI) is a term coined by John McCarthy, a professor at MIT. The field of AI is all about creating computer programs that can perform tasks that require high-level thinking, like perceptual learning, memory organization, and critical reasoning. The discipline was born in the summer of 1956 at a conference at Dartmouth College, funded by the Rockefeller Institute. Despite a promising start, the popularity of AI decreased in the early 1960s when machines had limited memory, and it was tough to use computer language.

However, some foundations laid in the field, such as solution trees to solve problems and information processing language (IPL), showed promising results. In 1956, the IPL helped to write the "Logic Theorist Machine" program that aimed to demonstrate mathematical theorems. Herbert Simon, a famous economist and sociologist, predicted that AI would beat humans at chess in the next ten years, which turned out to be true, but only after three decades.

The 1968 film "2001 Space Odyssey" raised important ethical questions about the potential dangers and benefits of AI. This popularized the theme, along with science fiction authors like Philip K. Dick, who wondered if machines would one day experience emotions. The advent of microprocessors in the 1970s led to the golden age of expert systems, like MIT's DENDRAL and Stanford University's MYCIN. These systems were programmed with an "inference engine" that mirrored human reasoning and provided highly expert answers. However, programming such knowledge was difficult and could result in a "black box" effect with unclear reasoning. By the end of the 1980s and early 1990s, AI development had slowed due to these challenges and the rise of faster and less expensive alternatives. The success of IBM's Deep Blue in defeating Garry Kasparov at chess in 1997 was symbolic but did not support the development of expert systems. Deep Blue relied on a brute force algorithm and was limited to the rules of chess, far from modeling the complexity of the world.

In 2010, the field of artificial intelligence (AI) experienced a new wave of development, thanks to two factors: the abundance of massive data and the discovery of the high efficiency of computer graphics card processors. Before this, it was difficult to access large volumes of data, and learning algorithms took weeks to process the entire sample.

However, the availability of massive data and powerful processors changed the game. Today, a simple Google search can yield millions of data points, and the computing power of these cards has enabled significant progress at a limited financial cost.

This new technological equipment has led to some significant public successes and boosted funding. IBM's IA, Watson, won the game against two Jeopardy champions in 2011. In 2012, Google's search lab, Google X, used AI to recognize cats in a video, and in 2016, AlphaGO, Google's AI specialized in Go games, beat the European and world champions. These successes were made possible by a complete paradigm shift from expert systems to an inductive approach, where computers discover rules alone by correlation and classification, based on a massive amount of data.

Among machine learning techniques, deep learning seems the most promising for many applications, including voice and image recognition. In 2003, a research program was launched to bring neural networks up to date, and experiments showed that this type of learning succeeded in halving the error rates for speech recognition. Similar results were achieved by deep learning teams in image recognition. However, there is still a long way to go to produce text understanding systems and conversational agents. While our smartphones can transcribe an instruction, they cannot fully contextualize it and analyze our intentions.

Types of AI

Artificial Intelligence can be classified into three categories based on their capabilities, which are Narrow AI, General AI, and Super AI.

Moreover, AI can be categorized into four types based on functionalities, which are Reactive Machines, Limited Theory, Theory of Mind, and Self-awareness.

Narrow AI

Narrow AI, also known as Weak AI, is a type of artificial intelligence that is designed to perform a single task or a narrow range of tasks. It focuses on a limited set of cognitive abilities and cannot perform beyond its programmed limitations. Narrow AI is prevalent in our daily lives, from voice assistants like Siri and Alexa to image recognition software and recommendation algorithms. The advancements in machine learning and deep learning have made it possible to develop more sophisticated narrow AI applications that can perform specific tasks with high accuracy and efficiency.

Artificial Intelligence (AI) systems can be categorized based on their functions. One of the primary forms of AI is called a reactive machine.

This type of AI does not store memories or use past experiences to determine future actions. Instead, it works solely with present data, perceiving the world and reacting to it. Reactive machines are designed to perform specific tasks and have no capabilities beyond those tasks.

An instance of a reactive gadget is IBM's Deep Blue, which famously defeated chess grandmaster Garry Kasparov. Deep Blue sees the chessboard pieces and reacts to them, but it cannot refer to any of its prior experiences or improve with practice. The machine can identify the pieces on a chessboard, understand how each piece moves, and predict possible next moves for both itself and its opponent. However, it ignores everything that happened before the present moment and looks only at the chessboard pieces as they currently stand to make its next move.

- The Theory of Mind

The Theory of Mind is a concept that refers to an advanced form of AI technology. It is designed to understand that people and objects in an environment can influence feelings and behaviors. This type of AI is expected to comprehend human emotions, thoughts, and sentiments. While significant progress has been made, it is still incomplete.

An example of Theory of Mind AI is Kismet, a robot head developed in the late 1990s by a Massachusetts Institute of Technology researcher. Kismet can mimic human emotions and recognize them, but it lacks the ability to follow gazes or convey attention to humans.

Another Example is Sophia, created by Hanson Robotics. Sophia can see through cameras in her eyes and recognize individuals, sustain eye contact, and follow faces. These are important advancements in Theory of Mind AI.

- Limited Memory

Limited Memory AI is a technology that utilizes past data to make decisions. It has a short-lived memory and is unable to add the acquired information to a library of experiences. However, it can use the past data for a specific period of time. The technology is particularly useful in the development of self-driving vehicles.

In the case of Limited Memory AI, self-driving vehicles are equipped with sensors that observe and gather information about the movement of other cars around them. This collected data is then added to the AI's existing data, such as lane markers and traffic lights. By combining the ongoing observations with the static data, the vehicle can make informed decisions such as when to change lanes, avoid cutting off another driver, or when to stop to prevent a collision.

Mitsubishi Electric is one of the companies that has been working on improving Limited Memory AI for the development of self-driving cars. The technology is constantly evolving, and with further advancements, it is expected to become more efficient and reliable in making accurate decisions while driving.

With the increasing demand for self-driving cars, Limited Memory AI is becoming an essential technology that makes driving safer and more convenient for people.

General AI

General AI, also known as strong AI, is a type of artificial intelligence that is capable of understanding and learning any intellectual task that a human being can.

Unlike narrow AI, which is programmed to perform specific tasks, General AI can apply its knowledge and skills in different contexts, similar to how humans learn and adapt to new situations. However, despite significant advancements in AI, researchers have not yet been able to achieve strong AI. To achieve this feat, they would need to find a way to make machines conscious and program a full cognitive ability set.

Recently, Microsoft invested Recently, Microsoft invested $1 billion in OpenAI, an organization dedicated to developing advanced AI technologies, including General AI. billion in OpenAI, an organization dedicated to developing advanced AI technologies, including General AI. Fujitsu, a Japanese tech company, has also made significant attempts at achieving strong AI with its K computer, which is one of the fastest supercomputers in the world. However, it still took nearly 40 minutes to simulate a single second of neural activity, making it challenging to determine when strong AI will be achieved.

Super AI

Super AI, on the other hand, is a hypothetical type of artificial intelligence that surpasses human intelligence and can perform any task better than a human. It is often associated with the concept of artificial super intelligence, where AI evolves to be so akin to human sentiments and experiences that it not only understands them however additionally conjures up emotions, needs, beliefs, and goals of its own. The existence of Super AI is still hypothetical, and some of its critical characteristics include thinking, solving puzzles, making judgments, and decisions on its own.

Significance OF AI

Artificial intelligence, or AI, can be defined in many ways, but the most important conversation revolves around what AI enables you to do. It can improve end-to-end efficiency by eliminating friction, improving analytics, and optimizing resource utilization across your organization, resulting in significant cost reductions. AI can also automate complex processes and minimize downtime by predicting maintenance needs, making it a valuable asset for any business.

One of the most significant benefits of AI is its ability to augment human intelligence. By providing rich logic and pattern prediction capabilities, AI can improve the quality, effectiveness, and creativity of employee decisions, resulting in improved accuracy and decision-making. This technology can help businesses uncover gaps and opportunities in the market more quickly, allowing them to introduce new products, services, channels, and business models with a level of speed and quality that wasn't previously possible.

AI can also empower employees by taking care of mundane tasks, freeing up time for high-value tasks that can be more fulfilling and engaging. This can fundamentally change the way work is done and reinforce the role of people in driving growth. Additionally, using AI can unlock the incredible potential of talent with disabilities, while helping all workers thrive.

Another area where AI can be incredibly beneficial is in providing superior customer service. Continuous machine learning can provide a steady flow of 360-degree customer insights for hyper-rationalization. From 24/7 chat bots to faster help desk routing, businesses can use AI to curate information in real-time and provide high-touch experiences that drive growth, retention, and overall satisfaction.

It's important to note that your AI strategy is your business strategy. To maximize your return on AI investments, it's essential to identify your business priorities and determine how AI can help. With the right AI strategy in place, businesses can reap the benefits of this informativeness technology and stay ahead of the competition.

Applications of AI

Artificial intelligence (AI) is applied in various sectors for different purposes. One of the most common uses of AI is facial recognition, which is used in personal devices like phones, laptops, and PCs, as well as in high-security areas across different industries. Facial recognition uses face filters to detect and identify individuals for secure access. Another widely used AI application is the recommendation system, which is commonly used by e-commerce, entertainment, social media, and video-sharing platforms to provide customized recommendations to users and increase engagement.

Navigation is another sector that benefits from AI. MIT's research shows that GPS technology with Convolutional Neural Networks and Graph Neural Networks can provide accurate, timely, and detailed information to improve safety. This technology can automatically detect the number of lanes and road types behind obstructions on the road, making it easier for users to navigate. Uber and logistics companies use AI to improve operational efficiency, analyze road traffic, and optimize routes.

Robotics is also a field where AI applications are commonly used. Robots powered by AI are programmed to use real-time updates to sense obstacles in their path and pre-plan their journey accordingly. They are used for carrying goods in hospitals, factories, and warehouses, cleaning offices and large equipment, and inventory management.

In the human resource sector, AI is used to ease the hiring process. AI helps with blind hiring by examining applications based on specific parameters using machine learning software. AI drive systems can scan job candidates' profiles and resumes to provide recruiters with an understanding of the talent pool they must choose from.

AI finds diverse applications in the healthcare sector as well. Healthcare organizations use AI to build sophisticated machines that can detect diseases and identify cancer cells. AI can analyze chronic conditions with lab and medical data to ensure early diagnosis. AI makes use of the aggregate of historic information and scientific intelligence for the invention of latest drugs.

TRENDS IN AI

Artificial intelligence is rapidly evolving, and its applications are becoming increasingly widespread. Multimodal AI, for instance, is being used to improve the accuracy of diagnostic tools and extend the capabilities of design and coding. Additionally, it is opening up new avenues for learning by providing AI systems with fresh data to learn from.

Another exciting development in the world of AI is agentic AI. This shift from reactive to proactive AI means that AI agents can now understand their environment, set goals, and act to achieve those objectives without the need for human intervention. This has huge implications for industries ranging from healthcare to supply chain management, where proactive AI systems can improve efficiency and reduce costs.

Open source AI is another trend that is gaining momentum. It is publicly available and often free, enabling organizations and researchers to contribute to and build on existing code. This approach can encourage transparency and ethical development, as more eyes on the code can increase the likelihood of identifying biases, bugs, and security vulnerabilities. However, there are concerns about the misuse of open source AI to create disinformation and other harmful content.

Retrieval-augmented generation (RAG) is a fascinating technique that blends text generation with information retrieval to enhance the accuracy and relevance of AI-generated content. RAG enables LLMs to access external information, helping them produce more accurate and contextually aware responses. This is particularly useful in enterprise adoption, where inaccuracies could be catastrophic. Customized enterprise generative AI models are becoming increasingly popular as they cater to niche markets and user needs. They are more efficient and informative and can be built at a lower cost than building a new model from scratch. In addition, customized models can be built to handle sensitive data, improving privacy and security.

However, there is a shortage of AI and machine learning talent, particularly in the area of MLOps. Organizations need to build internal AI and machine learning capabilities and focus on diversity in AI initiatives to challenge biases and improve results.

Finally, the issue of shadow AI is becoming increasingly prevalent as AI becomes more accessible. Shadow AI is where AI is used within an organization without explicit approval or oversight from the IT department. It carries risks related to security, data privacy, and compliance, so it is essential for organizations to think through the appropriate control measures.

As we move into 2024 and beyond, these AI trends are set to shape the future of industries and society as a whole. It is an exciting time to be part of the AI revolution, and the possibilities for innovation and progress are limitless.

Challenges in AI

Artificial Intelligence (AI) is undoubtedly one of the most exciting fields of technology today, but it is not without its challenges. In this article, we will discuss the most common problems in AI and potential solutions for each issue.

The first challenge is computing power. Machine Learning and Deep Learning, the building blocks of AI, require an increasing number of cores and GPUs to function effectively. While cloud computing and parallel processing systems are available, they come at a price that not everyone can afford.

The second challenge is the trust deficit. Many people are unaware of how deep learning models predict outputs, which can lead to skepticism and mistrust of AI. This lack of understanding can hinder the adoption of AI in various industries.

The third challenge is limited knowledge. While there are many potential use cases for AI, few people outside of technology enthusiasts, college students, and researchers are aware of its potential. This lack of awareness can result in missed opportunities for small and medium-sized enterprises (SMEs) to increase production, manage resources, and understand consumer behavior.The fourth challenge is achieving human-level performance.

While some AI services boast above 90% accuracy, humans can still outperform them.

Achieving human-level performance requires significant finetuning, hyperparameter optimization, large datasets, and a well-defined algorithm, all of which are challenging to achieve.

The fifth challenge is data privacy and security. The availability of data is essential for training AI models, but there is a risk that this data can be used for malicious purposes. Data breaches can result in the theft of sensitive personal information, which raises ethical concerns.

The sixth challenge is the bias problem. AI systems are only as good as the data they are trained on. Biases in the data can result in biased AI models, which can lead to unintended consequences.

Finally, the seventh challenge is data scarcity. Major companies are facing charges regarding the unethical use of user data, and some countries are imposing stringent IT rules to restrict the flow of data. With biased information, the entire AI system could become flawed.

To overcome these challenges, companies and organizations must prioritize transparency, accountability, and ethical considerations when developing and deploying AI. Collaboration between different stakeholders, including researchers, policymakers, and industry experts, is essential to ensure that AI is developed and used in a responsible and ethical manner.

Opportunities in AI

The demand for artificial intelligence (AI) careers has been increasing recently due to the increased demand in various industries. The potential of AI creating a plethora of new jobs is justified, and a career in AI appears to be more promising than any other job available today. Before you dive into AI career opportunities, it's essential to understand what AI is and what AI careers you can pursue.

Let me hint briefly on the career opportunities that you should know in the year 2024. These career opportunities include Big Data Engineer, Data Scientist, Machine Learning Engineer, Business Intelligence Developer, Research Scientist, Product Manager, AI Engineer, AI Data Analyst, Robotics Scientist, and NLP Engineer.

A Big Data Engineer's primary responsibility is to generate and effectively manage big data for an organization. They must also obtain reliable results from big data. Being a Big Data Engineer pays well, and the average salary is Rs. 8.7 LPA.

Data Scientists assist in collecting relevant data from multiple sources to analyze it and draw constructive conclusions that apply to a wide range of business-related issues. They make predictions based on data patterns, and the average salary starts at Rs. 8.7 LPA.

Machine Learning (ML) is widely recognized as a subset of AI. Machine Learning Engineers are responsible for developing and maintaining self-running software that supports machine learning initiatives. Their annual salary is approximately Rs. 7.34 LPA.

The role of a Business Intelligence Developer is to identify various business trends by analyzing large amounts of data. They make contributions to a company's income by way of means of planning, developing, and nurturing commercial enterprise intelligence solutions. A Business Intelligence Developer's annual salary is Rs. 6.6 LPA.

Research Scientists conduct large studies on system getting to know and its applications. They must know applied mathematics, statistics, deep learning, and machine learning. A Research Scientist's annual salary is Rs. 7.8 LPA.

In the field of AI, a Product Manager's role is to solve difficult problems by strategically collecting data. They must be able to identify issues that are impeding business operations, collect related data sets to help with data interpretation, and estimate the business implications of the results. A Product Manager's annual salary is approximately Rs. 17.5 LPA.

AI Engineers are problem solvers who develop, test, and apply AI models. They use system mastering algorithms and an information of neural networks to create beneficial AI models. An AI Engineer's average salary is around Rs. 6 LPA.

The primary responsibility of an AI Data Analyst is to perform data cleaning, data mining, and data interpretation. Data Scientists earn an average of Rs. 4.7 LPA. Robotics Scientists program machines for major industries that rely on robotics to fulfill their tasks efficiently. They require a master's degree in computer science, robotics, or engineering, and their annual salary is approximately Rs. 4.48 LPA.

Natural Language Processing (NLP) Engineers specialize in human language, including both spoken and written information. They work on speech recognition, voice assistants, document processing, and more. Organizations expect NLP Engineers to have a specialized degree in computational linguistics and may also consider candidates with degrees in mathematics, computer science, or statistics.

Regulating AI

Regulating artificial intelligence (AI) systems is a process that entails creating frameworks, policies, and guidelines that govern the development, deployment, and use of AI technology. The goal is to ensure that AI technologies are developed ethically, responsibly, and in a manner that aligns with societal values and safety considerations. There are several key aspects of AI regulation that are important to consider.

Firstly, ethical guidelines aim to establish guidelines that ensure AI systems respect human rights, privacy, fairness, transparency, and accountability. These guidelines are meant to address issues like bias mitigation, data privacy, and the responsible use of AI in decision-making processes.

Secondly, safety and reliability standards are crucial, particularly in critical domains like healthcare, autonomous vehicles, and aerospace. These standards focus on ensuring reliability, robustness, and resilience to prevent AI failures that could lead to harm.

Thirdly, transparency and explainability are essential, requiring AI systems to make their decisions and functioning understandable and interpretable. Regulations may necessitate explanations for AI decisions, especially in applications impacting individuals' lives, such as lending or hiring decisions.

Fourthly, data governance and privacy are other critical aspects of regulating AI. This involves addressing data governance to ensure the responsible collection, storage, and use of data. Privacy regulations, like the GDPR (General Data Protection Regulation) in the European Union, impose restrictions on data usage and empower individuals with rights over their data.

Fifthly, human oversight and accountability are also important. Guidelines may mandate human oversight in critical AI systems to ensure human accountability and intervention when necessary. This includes defining responsibilities and liabilities for the outcomes of AI systems.

Sixthly, international collaboration and standardization efforts are essential, given that AI operates globally. Collaborative initiatives aid in developing common principles and guidelines for responsible AI development and deployment.

Seventh, regulation specific to AI applications is necessary in certain sectors, such as healthcare, finance, and autonomous vehicles, to address unique challenges and risks associated with AI applications in those domains.

Lastly, continuous monitoring and adaptation are required given the rapid evolution of AI. Regulatory frameworks must be adaptable and regularly updated to keep pace with technological advancements and emerging risks.

Developing and implementing regulations for AI involves collaboration among governments, industry experts, researchers, and policymakers. Striking a balance between fostering innovation and safeguarding societal well-being remains a crucial challenge in the dynamic landscape of AI regulation.

CHAPTER 2

- Machine Learning

Welcome to the world of Machine Learning, where the impossible becomes possible, and the unimaginable becomes reality. In this intricate web of technological evolution, Machine Learning stands out as the embodiment of innovation and transformation. It's a field that has redefined the landscape of modern technology and continues to shape our future.

At its core, Machine Learning is not just a tool; it's a paradigm shift—an art of enabling computers to learn and improve from experience without being explicitly programmed for each task. It's the force behind predictive logic, recommendation systems, autonomous vehicles, and medical diagnostics—a force that shapes efficiency, innovation, and human experiences.

But what makes Machine Learning so captivating? It's the ability of machines to evolve, learn, and adapt without explicit programming, where algorithms decipher patterns hidden within colossal data sets, and where the impossible becomes probable, courtesy of computational prowess. This is the captivating realm of Machine Learning.

To embark on this journey, we delve into the fundamental pillars of Machine Learning, where algorithms serve as the architects of this computational evolution. Through supervised, unsupervised, and reinforcement learning, machines uncover elusive patterns, turning raw data into actionable insights.

But what breathes life into these algorithms? Enter the world of neural networks—an intricate web of interconnected nodes inspired by the human brain. These neural networks, with their layers and neurons, unravel the mysteries hidden within data, providing machines with the ability to perceive, comprehend, and make decisions, albeit in a digital realm.

However, the magic of Machine Learning extends beyond comprehension; it transcends industries, revolutionizing healthcare, finance, transportation, and more. It's the force behind efficiency, innovation, and human experiences.

As we bask in the brilliance of Machine Learning, we acknowledge the ethical quandaries and societal responsibilities that accompany this technological leap. Questions of bias, privacy, and accountability echo in the corridors of this digital evolution, demanding ethical frameworks and conscious governance.

Join us on this expedition into the realm where algorithms learn, adapt, and evolve—a journey where the intersection of data, algorithms, and computation redefines what's achievable. Here, within the captivating world of Machine Learning, lies the informativeness power that reshapes industries, empowers innovation, and charts the trajectory of our future.

Buckle up as we navigate the corridors of this enchanting domain, unraveling the mysteries and unveiling the magic of computational learning—ushering in an era where machines not only compute but also comprehend, creating a tapestry where imagination meets innovation.

Types of Machine Learning Algorithm

Supervised Learning

Supervised learning is a machine learning technique that involves training algorithms using labeled data. In supervised learning, the algorithm receives input data and corresponding correct output labels during the training process. The objective is to train the algorithm to predict accurate labels for new, unseen data.

Supervised learning can be used for different types of tasks, including classification, regression, and time series forecasting. Examples of supervised gaining knowledge of algorithms encompass choice trees, aid vector machines, random forests, and Naive Bayes.

Supervised learning has been widely used in various domains, including healthcare, finance, marketing, and image recognition, to make predictions and gain valuable insights from data.

Unsupervised Learning.

Unsupervised learning is a machine learning approach that involves exploring patterns, relationships, or structures within data without predefined output labels. In unsupervised learning, algorithms analyze unlabeled data to discover hidden insights and group similar data points together. Common unsupervised learning techniques include clustering algorithms like K-means and hierarchical clustering, as well as dimensional reduction methods like PCA and t-SNE.

Unsupervised learning has been used in various domains, including data mining, anomaly detection, and recommendation systems.

Semi-Supervised Learning

Semi-supervised studying is a hybrid system studying method that mixes categorized and unlabeled records for training. In semi-supervised learning, the algorithm leverages the limited labeled data and a larger set of unlabeled data to improve the learning process. The idea is that the unlabeled data provide additional information and context to enhance the model's understanding and performance.

Semi-supervised learning is particularly useful when acquiring labeled data is expensive or time-consuming. It can be applied to various tasks, such as classification, regression, and anomaly detection, allowing models to make more accurate predictions and generalize better in real-world scenarios.

Reinforcement Learning

Reinforcement learning is a machine learning approach that involves training algorithms to make decisions in an environment to maximize a reward or minimize a penalty. In reinforcement learning, the algorithm interacts with the environment by taking actions and observing the results. The objective is to learn the optimal policy that maximizes the expected cumulative reward.

Reinforcement learning has been used in various domains, including robotics, game playing, and recommendation systems.

Machine learning algorithms are classified into four main types: supervised learning, unsupervised learning, semi-supervised learning, and reinforcement learning. Each type of machine learning algorithm has its strengths and weaknesses and can be used for different types of tasks. Understanding the different types of machine learning algorithms is essential for developing machine learning models that can make accurate predictions and gain valuable insights from data.

List of Popular Learning Machine Algorithms.

Linear Regression

Linear Regression is a machine learning technique that involves arranging random logs of wood in increasing order of their weight without actually weighing them. Instead, one has to guess the weight by looking at the height and girth of the logs and creating a combination of these visible parameters.

The process involves establishing a relationship between independent and dependent variables by fitting them to a line, also known as the regression line. This line is represented by a linear equation $Y = a * X + b$, where Y is the dependent variable, a is the slope, X is the independent variable, and b is the intercept. The coefficients a and b are derived by minimizing the sum of the squared difference of distance between data points and the regression line.

Logistic Regression

Logistic Regression is a machine learning technique that is widely used to estimate discrete values, usually binary values such as 0 or 1, based on a set of independent variables. It involves fitting the data to a logit function to predict the probability of an event. This regression type is also known as logit regression. To enhance the accuracy of logistic regression models, different methods can be used, such as including interaction terms, eliminating features, applying regularization techniques, and using non-linear models.

Decision Tree

Decision Tree is a well-known algorithm in the machine learning community. It is used to classify problems and works well with both categorical and continuous data. The algorithm divides the population into two or more homogeneous sets based on the most significant attributes, making it easier to classify the data.

Support Vector Machine (SVM)

Support Vector Machine (SVM) is another popular algorithm used for classification. It plots raw data as points in an n-dimensional space and uses classifiers to split the data and plot them on a graph. SVM is widely used in the industry for solving classification problems.

Naive Bayes

Naive Bayes is a classification algorithm that assumes that the presence of a particular feature in a class is unrelated to the presence of any other feature. It is easy to build and useful for massive datasets.

Nearest Neighbors (KNN)

K-Nearest Neighbors (KNN) is a simple yet effective algorithm for classification and regression problems. It shops all to be had instances and classifies any new instances via way of means of taking a majority vote of its okay neighbors. The case is then assigned to the elegance with which it has the maximum in common.

K-Means

K-Means is an unsupervised learning algorithm that solves clustering problems. Data sets are classified into a particular number of clusters in such a way that all the data points within a cluster are homogenous and heterogeneous from the data in other clusters.

Random Forest

Random Forest is a collective of decision trees that is used for classification. Each tree is classified, and the tree "votes" for the class. The woodland chooses the type having the maximum votes over all of the timber withinside the woodland.

.

As the amount of data being analyzed by organizations increases, it's becoming increasingly difficult to identify significant patterns and variables amidst the vast amounts of raw data. That's where dimensionality reduction algorithms like Decision Tree, Factor Analysis, Missing Value Ratio, and Random Forest come in handy. They help data scientists extract relevant details from massive amounts of data.

But what about when you need to make predictions with high accuracy? That's where Gradient Boosting Algorithm and AdaBoosting Algorithm come in. These powerful boosting algorithms combine the predictive power of several base estimators to improve robustness. By combining multiple weak or average predictors to build a strong predictor, they can achieve accurate outcomes. In fact, these algorithms are so effective that they're often used in data science competitions like Kaggle, AV Hackathon, and CrowdAnalytix. So if you want to stay ahead of the game in data science, make sure to use these algorithms along with Python and R codes.

CHAPTER 3

- **Natural Language Processing**

Natural Language Processing (NLP) is a mind-blowing technology that has made it possible for computers to understand, generate, and manipulate human language. It's like having a super-intelligent language translator by your side! With NLP, you can communicate with computers in a natural way, without having to learn a programming language. It's like having a conversation with a human being, only better! Plus, it's not just limited to written text - you can use your voice too!

NLP can help you in various ways, from simple tasks like searching the web, checking your emails, and filtering out spam messages, to more complex tasks like summarizing large amounts of information and understanding customer feedback. With NLP, you can interact with technology like never before and get things done more efficiently.

NLP is the result of years of research and development, and is powered by machine learning algorithms that learn and generalize from vast amounts of data. This means that NLP is becoming more and more accurate and effective, making it an essential tool for businesses and individuals alike. Imagine being able to analyze customer feedback and sentiment in an instant, or being able to summarize complex information into easy-to-understand text with just a few clicks. With NLP, it's all possible!

In conclusion, NLP is a revolutionary technology that has opened up a whole new world of possibilities. It's like having a personal language assistant that can help you with anything you need! Whether you're a student, a professional, or just someone who wants to interact with technology in a more natural way, NLP is the technology of the future.

Applications of NLP

Natural Language Processing (NLP) is a modern technology that has a wide range of applications for businesses. It can automate routine tasks, improve search, and provide valuable market insights. One of the most significant benefits of NLP is the ability to use chatbots and digital assistants to handle routine requests, freeing up employees to focus on more challenging tasks. NLP-powered chatbots can handle customer queries, provide support, and even assist with transactions, making it easier for businesses to provide round-the-clock assistance to their customers.

NLP can also improve search by analyzing the context, synonyms, and morphological variations of search queries. This technology can help businesses retrieve relevant documents and information faster, which is essential in today's fast-paced business environment. Additionally, NLP can be used to analyze customer reviews and social media comments to provide valuable market insights. With sentiment analysis, businesses can understand their customers' needs and preferences, which can help them improve their products and services.

Moreover, NLP can help moderate content by analyzing the tone and intent of comments. This technology can flag inappropriate comments and maintain quality and civility in online forums and social media platforms. Additionally, NLP can help businesses analyze and organize large amounts of data, which can be overwhelming and time-consuming. By using NLP-powered tools, businesses can quickly extract meaningful insights from data, which can help them make informed decisions and stay ahead of the competition.

Overall, NLP is a powerful technology that can provide significant benefits to businesses across various sectors. By automating routine tasks, improving search, and providing valuable market insights, NLP can help businesses become more efficient, effective, and customer-focused.

Industries involved in Natural Language Processing

Natural Language Processing (NLP) is proving to be a game-changer for many industries by simplifying and automating various business processes that involve huge amounts of unstructured text like surveys, emails, social media conversations, and more. NLP enables businesses to analyze their data more effectively, leading to better decision-making. Below are some practical applications of NLP:

Healthcare: As healthcare systems transition to electronic medical records, they are faced with large volumes of unstructured data. NLP can help analyze and gain new insights into health records.

Legal: Lawyers often spend hours examining large collections of documents to prepare for a case. NLP technology can automate the legal discovery process, effectively reducing both time and human error by sifting through large volumes of documents.

Finance: In the fast-paced financial world, any competitive advantage is critical. Traders use NLP technology to automatically mine information from corporate documents and news releases to extract relevant information for their portfolios and trading decisions.

Customer service: Many large companies are using virtual assistants or chatbots to answer basic customer inquiries and information requests, passing on complex questions to humans when necessary.

Insurance: Large insurance companies are using NLP to streamline their operations by sifting through documents and reports related to claims.

NLP Technology Overview

Natural Language Processing (NLP) has become an essential field in the world of artificial intelligence. Machine learning plays a crucial role in NLP, helping to analyze data and make predictions using training data. This data is used to create machine learning models that can perform specific tasks, such as sentiment analysis. These models are called document classification models and can classify documents into topics like sports, finance, or politics.

NLP also uses entity recognition models that identify and classify entities in a document by predicting whether each word is part of an entity mention and the type of entity involved. Additionally, sequence labeling models generate a label for each word in the input, while sequence-to-sequence models produce a program or sequence as output.

Deep learning is a widely used machine learning technique in NLP, and deep neural network models are extensively used because of their complexity. However, transfer learning makes it possible to further train a trained neural network model to achieve a new task with less data and compute effort. Furthermore, pretrained models are available and can be fine-tuned for different target tasks. This ecosystem of providers has simplified the deployment of deep learning models across various industries and use cases.

Sample of NLP Preprocessing Techniques

Here are some commonly used NLP preprocessing techniques:

- Tokenization: This involves breaking down a text into a sequence of tokens, such as words or characters. These tokens are treated as atomic units in later processing.

- Bag-of-words models: These models treat documents as unordered collections of tokens or words, ignoring their order. They are often used for efficiency reasons in information retrieval tasks.

- Stop word removal: Stop words, such as "the" or "a," are frequently occurring words that are ignored in later processing to reduce processing time and storage.

- Stemming and lemmatization: These techniques map words to their stem forms to reduce the number of unique tokens. They are crucial preprocessing steps in traditional NLP models, but are not required in deep learning models.

- Part-of-speech (PoS) tagging and syntactic parsing: These involve labeling each word with its part of speech and identifying how words combine to form phrases and sentences. While essential in traditional NLP models, they are not widely used in deep learning NLP as neural networks can learn these regularities from their training data.

One of the primary programming languages used for NLP is Python, thanks to the availability of NLP libraries and toolkits in this language. Python's interactive development environment makes it easy to develop and test new code, making it a popular choice for NLP projects. However, for processing large amounts of data, C++ and Java are often preferred because they can support more efficient code.

Here are some examples of popular NLP libraries that developers can use to build NLP applications:

- TensorFlow and PyTorch: These are the 2 most famous deep mastering toolkits. They are freely available for research and commercial purposes, and their primary language is Python. They come with large libraries of pre-built components, so even very sophisticated deep learning NLP models often only require plugging these components together. They additionally assist high-overall performance computing infrastructure, along with clusters of machines with graphical processor unit (GPU) accelerators. They have excellent documentation and tutorials.

- AllenNLP: This is a library of high-stage NLP components (for example, easy chatbots) applied in PyTorch and Python. The documentation is excellent.

- HuggingFace: This organization distributes masses of various pretrained deep studying NLP models, in addition to a plug-and-play software program toolkit in TensorFlow and PyTorch that allows builders to swiftly compare how properly distinctive pretrained fashions carry out on their unique tasks.

- Spark NLP: Spark NLP is an open-source text processing library for advanced NLP for the Python, Java, and Scala programming languages. Its intention is to offer an software programming interface (API) for herbal language processing pipelines. It gives pre trained neural community models, pipelines, and embeddings, in addition to guide for schooling custom models.

- SpaCy NLP: SpaCy is a free, open-supply library for superior NLP in

Python, and it's mainly designed to assist construct packages that may system and apprehend massive volumes of text. SpaCy is thought to be pretty intuitive and might manage most of the obligations wanted in not unusual NLP projects.

Oracle Cloud Infrastructure is committed to providing on-premises performance with performance-optimized compute shapes and tools for NLP. The cloud infrastructure offers an array of GPU shapes that can be deployed in minutes to begin experimenting with NLP.

In summary, NLP is an exciting field that has a significant impact on the development of artificial intelligence. The programming languages and libraries used for NLP are varied and depend on the nature of the project. Python is the most popular language used for NLP due to its interactivity, while C++ and Java are preferred for processing large amounts of data. The availability of libraries such as TensorFlow, PyTorch, AllenNLP, HuggingFace, Spark NLP, and SpaCy

NLP have made it easier for developers to build and deploy NLP applications. The demand for growth in natural language processing will continue to increase as human interfaces with computers continue to evolve.

CHAPTER 4

- Computer Vision

Computer Vision is a fascinating field that harmonizes with human perception by empowering machines to not only see but also interpret, analyze, and comprehend visual information. It unlocks a realm of possibilities that go beyond human vision and redefines how we interact with the visual tapestry of our world. This cognitive endeavor is not just a field of study, but a discipline that allows machines to acquire, process, and interpret visual data much like the human eye and brain.

To understand Computer Vision, we must recognize that pixels on a screen are the building blocks of a digital canvas waiting to be deciphered. Machines are capable of discerning shapes, objects, patterns, and movements within these pixelated landscapes through the lens of algorithms and computational models. The algorithms serve as the eyes and brain for machines, enabling them to understand the visual world with an ever-increasing depth of perception.

Neural networks are the key to this understanding, as they are sophisticated systems inspired by the human brain's architecture. Convolutional Neural Networks (CNNs) are the stalwarts of Computer Vision, capable of learning and extracting hierarchical features from images, much like how the human visual system perceives edges, shapes, and textures.

Computer Vision extends far beyond static images and encompasses the dynamic realm of video analysis. This expansion broadens the scope of applications, ranging from surveillance and robotics to augmented reality and healthcare diagnostics. Witness how it drives innovation across industries—enhancing medical imaging, revolutionizing autonomous vehicles, powering facial recognition systems, and enabling augmented reality experiences that blur the boundaries between the digital and physical worlds.

However, as with any technological leap, the ethical and societal dimensions of Computer Vision require our attention. We must consider ethical considerations surrounding privacy, bias, and accountability in the development and deployment of vision-based technologies. Join us on this captivating odyssey, a journey into a world where machines not only perceive but also comprehend, interpret, and augment our visual experiences. Here, within the boundless horizon of Computer Vision, lies the transformative power that reshapes industries, redefines human-machine interaction, and illuminates the tapestry of our digital existence. Get ready to witness the symphony of sight and cognition interwoven into the fabric of machines, ushering in an era where pixels are not just points of light but gateways to understanding.

Grovety faced several challenges during their project to enhance the efficiency of wildlife camera traps through artificial intelligence. However, they successfully overcome these challenges through extensive research, development, and collaboration. They selected the Alif Semiconductor E7 processor, which is designed for power efficiency, long battery life, high computation, and machine learning capability.

Grovety leveraged the open-source machine learning compiler framework Apache TVM for fine-tuning their ML models, and made significant contributions to TVM for Arm Ethos-U55 integration. They also used the open-source Vela compiler to optimize their neural network model, which runs on the Ethos-U NPU. Grovety's solution aimed to significantly reduce false positives, improve efficiency, and extend battery life, which will help conservationists to monitor and protect endangered species more effectively. Overall, Grovety's expertise in software design, AI integration, and dedication to addressing real-world challenges in wildlife conservation, combined with Arm's advanced processors, yielded a groundbreaking prototype for wildlife camera traps.

Image Processing

AI image processing is a revolutionary technology that combines the power of artificial intelligence and computer vision to enable machines to interpret and manipulate visual data much like humans. It's a remarkable dance between algorithms and pixels, where machines "see" images and glean insights that elude the human eye. This technology has the potential to transform various industries, from precision agriculture to medical imaging, autonomous vehicles, and retail. By leveraging AI image processing, organizations can automate complex tasks and gain valuable insights that help them make data-driven decisions.

To get started with AI image processing, all you need is a machine learning image processing problem and a team of experts who can help you apply this technology to your specific use case.

AI image processing has the potential to bring about significant changes in various industries. For instance, precision agriculture is using drones and AI to monitor crop health with unprecedented detail. The film industry is also benefiting from stunning visual effects crafted by AI algorithms. In the medical field, AI image processing algorithms are scrutinizing medical scans to identify anomalies that are invisible to the human eye. Autonomous vehicles are also leveraging this technology to navigate bustling streets, detecting pedestrians and obstacles in real-time.

Even retailers are optimizing store layouts based on customer movement patterns tracked by AI cameras. The crux of this significance lies in the ability to extract valuable information from images, revolutionize decision-making, automate complex tasks, and explore more creative avenues.

If you're looking to automate repetitive manual tasks, you might want to check out Nanonets workflow-based document processing software. This tool allows you to extract data from invoices, identity cards, or any document on autopilot. If you're interested, you can even schedule a demo to see how it works.

Here's a simplified overview of how AI image processing works:

Data Collection and Preprocessing

The process begins with collecting a large data set of labeled images relevant to the task, such as object recognition or image classification. These images are preprocessed, which may involve resizing, normalization, and data augmentation to ensure consistency and improve model performance, Feature Extraction.

Convolutional Neural Networks (CNNs), a type of deep learning architecture, are commonly used to extract hierarchical features from images. CNNs automatically learn and extract patterns like edges, textures, and more complex features. They consist of layers with learnable filters (kernels) that detect these patterns.

Model Training

The Preprocessing images are fed into the CNN model for training. During training, the model adjusts its internal weights and biases based on the differences between its predictions and the actual labels in the training data. Back propagation and optimization algorithms (e.g., stochastic gradient descent) are used to interactively update the model's parameters to minimize prediction errors.

Validation and Fine-Tuning

A separate validation data set monitors the model's performance during training and prevents over-fitting (when the model memorizes training data but performs poorly on new data). Hyper parameters (e.g., learning rate) may be adjusted to fine-tune the model's performance.

Inference and Application

Once trained, the version is prepared for inference, which tactics new, unseen pics to make predictions. The AI image processing model analyzes the features of the input image and produces predictions or outputs based on its training.

Post-Processing and Visualization

Post-processing strategies can be carried out relying on the challenge to refine the model's outputs. For example, object detection models might use non-maximum suppression to eliminate duplicate detection. The processed snap shots or outputs may be visualized or similarly applied in numerous applications, which include scientific diagnosis, independent vehicles, artwork generation, and more continuous Learning and Improvement, AI photograph processing fashions may be constantly progressed via a cycle of retraining with new records and fine-tuning primarily based totally on consumer remarks and overall performance evaluation.

The success of AI image processing depends on the availability of high-quality labeled data, the design of appropriate neural network architectures, and the effective tuning of hyper parameters. If you want to automate repetitive manual tasks, you can check out Nanonets workflow-based document processing software, which extracts data from invoices, identity cards, or any document on autopilot.

Challenges in AI Image Processing

AI image processing technology has been advancing rapidly in recent years, and there are several challenges that need to be addressed to improve its effectiveness. One of the major concerns is the privacy and security of the massive amounts of data used in training these models.

Handling sensitive visual information like medical images or surveillance footage requires robust measures to prevent unauthorized access and misuse.

Another challenge is the possibility of bias in AI models, which may lead to unfair outcomes, especially when decisions impact individuals or communities. It is crucial to strive for fairness and minimize any biases present in the training data. Addressing this challenge requires careful attention to data collection and Preprocessing, model architecture and training, and post-processing techniques.

Ensuring that AI models perform reliably across different scenarios and environments is another challenge. For instance, models need to be robust enough to handle variations in lighting, weather, and other real-world conditions. This challenge requires the development of robust model architectures and training processes that can handle these variations.

Finally, understanding why a model makes a certain prediction remains a challenge. While AI image processing can deliver impressive results, explaining complex decisions made by deep neural networks requires ongoing research. Addressing this challenge requires the development of methods to interpret and visualize the internal workings of AI models, making it easier for users to understand how the model arrived at a particular prediction.

Trends in AI Image Processing

The field of AI image processing is constantly evolving and has many current trends that are addressing various challenges. One such trend is Explainable AI (XAI), which aims to provide insights into how AI models arrive at their decisions. This approach can make the decision-making process more understandable and accountable, particularly as AI systems become more complex.

Another trend is Few-Shot and Zero-Shot Learning, which enables AI models to generalize from very limited examples. Traditional machine learning often requires large amounts of labeled data for training, but Few-Shot and Zero-Shot Learning can mimic human-like learning and reduce the need for extensive labeling.

Advanced Image Manipulation Techniques is another trend that involves generating highly realistic images and videos. However, this trend also raises concerns about misuse, such as deepfake creation. As a result, countermeasures and detection techniques are emerging to address these challenges.

Semi-Supervised and Self-Supervised Learning are also gaining popularity. These approaches aim to reduce the reliance on fully labeled datasets for training, making AI image processing more accessible and efficient.

Continual Learning is yet another trend, which enables AI systems to adapt and learn incrementally instead of training models from scratch each time new data becomes available. This approach is particularly useful for tasks that involve evolving visual contexts.

Neuro Symbolic AI combines the power of neural networks with symbolic reasoning to enhance interpretability and enable more structured, human-understandable representations in AI image processing models.

Finally, Meta-Learning is a trend that involves training AI models to quickly adapt to new tasks with minimal data, which could lead to more efficient and adaptable image-processing solutions.

If you have a machine learning image processing problem, leveraging these trends in AI image processing could offer some useful insights and solutions.

Benefits

AI-powered image processing offers a wide range of benefits to businesses and individuals alike. One of the most significant advantages is automation and efficiency, as AI can automate repetitive and time-consuming tasks while allowing employees to focus on higher-value tasks and decision-making. AI is also capable of analyzing and interpreting images much faster than humans, making it easily scalable and capable of handling large volumes of images. This leads to faster speed and scalability, reducing the time and resources required for image processing.

Another benefit of AI image processing is the ability to extract valuable information and insights from images, which can be used for trend analysis, forecasting, and informed decision-making. Additionally, AI image processing can improve the customer experience by enabling advanced visual search capabilities, recommending products based on image analysis, and enhancing image-based user interfaces. It can also provide personalized recommendations, content, and experiences by analyzing user-generated images and data.

AI image processing can also lead to cost savings by automating tasks and reducing manual intervention. It can also perform complex analysis on images that might be challenging for humans, such as identifying patterns in medical images, detecting anomalies in manufacturing processes, or predicting equipment failures based on visual data. In autonomous vehicles, AI image processing is crucial in enabling real-time decision-making by rapidly interpreting the environment and making split-second choices to ensure safety and optimal performance.

AI image processing can also aid researchers and scientists in fields like astronomy, biology, and geology by helping analyze and interpret vast amounts of visual data. It can also be used to develop assistive technologies that make visual information accessible to people with disabilities, enhancing inclusivity. Finally, AI algorithms can achieve high levels of accuracy in image analysis and interpretation, minimizing the risk of human errors that often occur during manual processing.

Object Detection and Recognition

When it comes to AI image recognition software for enterprises, there are many options available in the market. While one option is to use AI image-processing Python libraries to create a custom solution, this approach can be quite resource-intensive and time-consuming. Alternatively, you can choose to leverage a specialized and established AI image recognition platform like Nanonets that comes with various benefits.

Nanonets offers an intuitive interface that is easy to use, driving highly accurate and rapid batch processing. With just a few effortless clicks, you can automatically extract data from any image, making it a convenient solution for organizations with large volumes of images to process. Moreover, Nanonets is a versatile solution capable of seamlessly ingesting documents from diverse channels, effectively becoming a centralized processing nucleus for all your document needs.

Despite being a cloud-based solution, Nanonets takes security very seriously. It has various certifications and compliance with security standards like SOC 2 Type 2 and HIPAA, ensuring that your data remains secure. This feature is particularly important for organizations that have to deal with sensitive data.

On the other hand, object detection is a computer vision task that involves identifying and locating specific objects within an image or video. Unlike simple object classification, object detection goes beyond just recognizing objects. It also involves delineating their boundaries using bounding boxes, providing more detailed information about the presence, location, and extent of objects of interest in the visual data. This makes it a crucial tool for a wide range of applications, including surveillance, self-driving cars, and robotics, among others.

Object detection is a crucial mission in laptop imaginative and prescient that entails figuring out and localizing items in images. There are two primary approaches to object detection: traditional image processing techniques and deep learning-based methods.

Traditional image processing techniques rely on algorithms that are based on computer vision concepts like edge detection, color segmentation, and feature extraction. These methods typically do not require extensive training data and are unsupervised in nature. The famous device OpenCV is frequently hired for photograph processing tasks. However, those strategies may also lack the accuracy and robustness of deep learning-primarily based totally approaches.

On the other hand, modern object detection heavily relies on deep learning networks, particularly Convolutional Neural Networks (CNNs). These algorithms are part of a family of region-based CNNs, which divide an image into regions and apply a CNN to each one to extract features. Deep learning-based approaches leverage large datasets for training, enabling the models to learn intricate features and patterns from images. This has significantly improved the performance of object detection systems.

One popular object detection algorithm is the Convolutional Neural Network (CNN), which has revolutionized the field by enabling highly accurate and efficient object detection. CNNs achieve high accuracy but tend to be computationally expensive. Another improvement over the original R-CNN algorithm is Fast R-CNN, which uses a region of interest pooling layer to share computation for overlapping regions, resulting in faster and more efficient object detection without sacrificing accuracy.

Another popular object detection algorithm is YOLO (You Only Look Once), which treats detection as a regression problem and predicts bounding boxes and class probabilities directly from grid cells. This approach is fast and accurate, even on devices with limited resources.

YOLO takes a different approach from R-CNNs by dividing the input image into a grid and predicting bounding boxes and class probabilities directly from the grid cells. This results in fast and accurate object detection, even in real-time scenarios.

In conclusion, object detection techniques have come a long way, and deep learning-based methods, specifically CNNs, have revolutionized the field by enabling highly accurate and efficient object detection. While traditional image processing techniques have their advantages, CNNs and other deep learning-based approaches have significantly improved the performance of object detection systems, making them the go-to choice for most applications.

Understanding the difference between object recognition and object detection is crucial in the field of computer vision. Object recognition deals with the identification and classification of objects in digital images or videos. It focuses on determining the correct object category rather than precise localization. The goal of object detection, on the other hand, is to detect and locate objects of interest in an image or video, along with classifying them into different categories.

Object detection involves identifying the position and boundaries of objects in an image. This task is accomplished using state-of-the-art methods categorized into one-stage and two-stage methods. One-stage methods prioritize inference speed and include models such as YOLO, SSD, and RetinaNet. Two-stage methods prioritize detection accuracy and include models such as Faster R-CNN, Mask R-CNN, and Cascade R-CNN.

Both object recognition and object detection have diverse applications across various industries. Object detection finds its applications in autonomous driving, surveillance systems, robotics, and image-based search engines. It permits responsibilities together with automobile and pedestrian detection, item tracking, and interactive augmented truth experiences. Object recognition, on the other hand, is precious in picture categorization, content-primarily based totally picture retrieval, and visible seek engines.

The MSCOCO data set is the most popular benchmark for evaluating object detection models. Models are typically evaluated according to a Mean Average Precision metric. Understanding the basics of object recognition and object detection is essential in the field of computer vision, and it has many practical applications in various industries.

Deep Residual learning

The concept of deep residual learning is based on the idea that multiple nonlinear layers can asymptotically approximate complicated functions, so they can also asymptotically approximate the residual functions. Rather than expecting stacked layers to approximate H(x), the original function, we let these layers explicitly approximate a residual function F(x) := H(x) − x. The function F(x, {Wi}) represents the residual mapping to be learned, and the original function becomes F(x)+x. This reformulation is motivated by the degradation problem, which suggests that solvers may have difficulties in approximating identity mappings by multiple nonlinear layers.

With the residual learning reformulation, the solvers may simply drive the weights of the multiple nonlinear layers toward zero to approach identity mappings, making it easier for them to find the perturbations with reference to an identity mapping than to learn the function as a new one.

To implement residual learning, we adopt shortcut connections to every few stacked layers. The shortcut connections introduce neither extra parameter nor computation complexity, making it easy to compare plain/residual networks that simultaneously have the same number of parameters, depth, width, and computational cost. The dimensions of x and F must be equal, but if they are not, we can perform a linear projection by the shortcut connections to match the dimensions.

The form of the residual function F is flexible, and experiments in this Paper contain a feature F that has or 3 layers, at the same time as greater layers are possible. We have examined numerous plain/residual nets, and found steady phenomena. To provide instances for discussion, we have described two models for ImageNet. Our plain baselines are mainly inspired by the philosophy of

VGG nets, with convolutional layers that mostly have 3×3 filters, and down sampling directly by convolutional layers that have a stride of 2. The community ends with an international common pooling layer and a 1000-manner fully-linked layer with softmax. Our residual networks are based on the plain networks, with shortcut connections (solid line shortcuts when input and output dimensions are the same, and dotted line shortcuts when dimensions increase) that turn the network into its counterpart residual version.

The identity shortcuts can be directly used when the input and output are of the same dimensions, and when the dimensions increase, we consider two options: (A) the shortcut still performs identity mapping, with extra zero entries padded for increasing dimensions, or (B) the projection shortcut is used to match dimensions (done by 1×1 convolutions). We have evaluated our method on the ImageNet 2012 classification dataset and observed the degradation problem - the deeper 34-layer plain net has higher validation error than the shallower 18-layer plain net.

However, our residual nets showed better performance, with a 3.57% top-5 error rate for the 34-layer residual net compared to a 3.78% error rate for the 18-layer plain net. Our method also showed good generalization performance on other recognition tasks, such as object detection on PASCAL and MS COCO, with a 6.0% increase in COCO's standard metric (mAP@[.5, .95]), which is a 28% relative improvement. Based on deep residual nets, we won the 1st places in several tracks in ILSVRC & COCO 2015 competitions: ImageNet detection, ImageNet localization, COCO detection, and COCO segmentation.

Applications of Computer Vision

Computer vision is a branch of Artificial Intelligence (AI) that aims to enable machines to interpret, analyze, and comprehend visual data from the world around us. It teaches machines to process images or visual inputs at the pixel level and extract meaningful information from them. This technology functions as the "eyes" of an AI system, enabling machines to perceive and observe visual input.

Computer vision performs various tasks such as image classification, object detection, object tracking, and semantic segmentation. Image classification is a technique that categorizes an image, such as identifying whether an image contains a dog, a person's face, or a banana. Object detection uses image classification to recognize and locate objects in an image or video, while object tracking is a technique that tracks a particular object or multiple objects over time. Semantic segmentation is an advanced technique that classifies each pixel of an image to identify the objects it contains and determine the role of each pixel in the image.

Computer vision has a broad range of applications across various industries, including healthcare, transportation, manufacturing, agriculture, and retail. In healthcare, computer vision is used for X-ray analysis, cancer detection, and MRI analysis. For example, computer vision can automate the analysis of X-ray images and detect subtle patterns that are challenging to detect with the human eye. In transportation, computer vision is utilized in self-driving cars, pedestrian detection, and road condition monitoring. For instance, computer vision algorithms can detect and categorize objects such as traffic lights or road signs to build 3D maps or estimate motion.

In manufacturing, computer vision is used for defect detection, analyzing text and barcodes, fingerprint recognition, and 3D model building. For instance, computer vision can detect defects in products such as cracks in metals, paint defects, and poor prints, in sizes smaller than 0.05mm. In agriculture, computer vision is used for crop monitoring, automatic weeding, and plant disease detection. For instance, computer vision algorithms can automatically identify and locate crops in an image or video, and monitor their growth and health status.

In retail, computer vision is used for self-checkout, automatic replenishment, and people counting. For example, computer vision can enable customers to complete their transactions without requiring human staff, and automatically restock shelves by performing a complete inventory scan. Computer vision algorithms can also analyze in-store camera data and count the number of people entering and leaving the store.

Overall, computer vision is a rapidly growing field with numerous applications in different industries. The increasing demand for AI and machine learning technologies is driving the growth of computer vision, and it is expected to have an even greater impact on society in the future.

CHAPTER 5

- Robotics

Robotics is an interdisciplinary field that encompasses various branches of engineering and science, such as electronics engineering, mechanical engineering, and computer science. It involves the conception, design, operation, and manufacturing of robots, as well as sensory feedback and information processing. Robotics is a rapidly advancing field and is expected to revolutionize many aspects of our lives in the coming years. Various types of robots are being developed for different purposes, including industrial robots, service robots, medical robots, and military robots.

One of the primary advantages of robots is their ability to perform tasks that are too dangerous or difficult for humans to undertake. For instance, robots are commonly used in bomb detection and deactivation, oil rig inspection, and space exploration. Additionally, robots can be used to enhance human capabilities in various fields, such as medicine and education. For example, surgical robots have made minimally invasive surgeries possible, while educational robots are being developed to help young children learn in a fun and interactive way.

In recent years, there has been a significant focus on developing robots that can emulate human behavior and intelligence. These robots are designed to look and behave like humans, with the ability to walk, talk, and perform tasks like a human. These robots are often referred to as humanoid robots or androids. Humanoid robots are still in the early stages of development, but they have already been used in various fields, such as nursing and entertainment.

Many modern robots are inspired by nature and are known as bio-inspired robots. These robots mimic the functionality of natural systems, such as the movement of animals or the structure of plants. Bio-inspired robots have been developed for various purposes, including search and rescue missions, exploration, and environmental monitoring.

The field of robotics was coined by author Issac Asimov in a short story he wrote in the 1940s. In the story, Asimov proposed three guiding principles for the use of robotic machines, which have become known as Asimov's Three Laws of Robotics. These laws state that robots must not harm human beings, must follow instructions given by humans without violating the first law, and must protect themselves without breaking any other laws.

Types of Robots

Robots powered by AI represent a significant advancement in the field of robotics, enabling machines to not only possess cognitive abilities but also interact with and navigate the physical world. These robots are capable of performing tasks that range from simple automation to complex decision-making in real-world scenarios, redefining human-machine interaction.

At the core of AI-driven robots are algorithms that facilitate learning, reasoning, and decision-making. These robots are equipped with sensors that capture data from their environment, which is then fed into AI systems that process and interpret it. Through machine learning and other AI techniques, robots can adapt to their surroundings, make informed decisions, and perform tasks autonomously or collaboratively with humans.

The applications of AI-powered robots span various domains, including manufacturing, healthcare, logistics and transportation, customer service, agriculture, and exploration. In manufacturing, robots can enhance efficiency by performing intricate tasks, adapting to production line changes, and even learning from human demonstrations. In healthcare, robots can aid in surgeries, patient care, and rehabilitation, augmenting the capabilities of medical professionals. In logistics and transportation, robots can streamline processes, optimize routes, and deliveries, making them more efficient.

Moreover, AI-driven robots can interact with humans, displaying empathy and understanding through natural language processing and emotional recognition algorithms. These social robots can help humans in various domains, such as education, entertainment, and therapy.

However, the integration of AI in robots also raises ethical considerations, such as job displacement, safety, accountability, and biases within AI algorithms. Addressing these concerns becomes crucial to ensure the responsible and beneficial integration of AI-powered robots into society.

The future of robots in AI promises a landscape where machines seamlessly collaborate with humans, augmenting our capabilities, enhancing productivity, and challenging us to redefine our relationship with technology. As these robots evolve, they become not just tools but companions, collaborators, and catalysts for innovation, shaping a future where the synergy between artificial intelligence and physical embodiment transcends the boundaries of imagination.

Components

Robots are complex machines that rely on various components to operate and perform tasks. These components work together to provide the robot's functionality and performance in different environments.

One of the most important components of a robot is its sensors. These sensors act as the robot's senses, providing data about the environment. A camera is used for vision, LiDAR or radar is used for distance measurement, ultrasonic sensors are used for proximity, touch sensors are used for physical interaction, and so on. These sensors allow robots to perceive and react to their surroundings.

Another critical component is the actuators. Actuators are responsible for the robot's movement or manipulation. They include motors for driving wheels or joints, pneumatic systems for controlled movement, servos for precise motion in robotic arms, and grippers or end-effectors for handling objects. Without actuators, robots would not be able to perform physical tasks.

The control system is another essential component of a robot. It includes the hardware and software responsible for processing sensor data, making decisions, and controlling the robot's movements. The control system often involves microcontrollers, processors, or specialized controllers that interpret sensory information and execute appropriate actions.

Power supply is crucial for the robot's operation. Robots require a power source to operate, ranging from batteries for mobile robots to wired power sources for stationary or industrial robots. Efficient power management is crucial for prolonged operation and optimal performance.

Manipulators are components that enable precise and controlled movements to interact with objects or perform tasks. These components are critical in robots designed for manipulation or handling tasks. They include robotic arms and specialized tools or end-effectors.

The chassis or frame is the physical structure of the robot that houses and supports its components. The chassis can vary widely, from a simple frame for mobile robots to complex structures for industrial or humanoid robots.

Communication interfaces are essential for robots that require communication capabilities to interact with humans or other systems. These interfaces can include wireless communication modules (like Wi-Fi or Bluetooth), ports for connecting to external devices, or communication protocols for networked robots.

Perception and processing units consist of hardware components (like processors,

GPUs, or specialized chips) and software algorithms that enable the robot to interpret sensory data, make decisions, and learn from its environment using machine learning or AI techniques.

In conclusion, each of these components plays a critical role in the functionality and capabilities of robots. Depending on the robot's purpose and design, these components are integrated and optimized to achieve specific tasks efficiently and effectively.

Application of Robotics

Robotics has become a widely used technology in many fields and industries, bringing about significant changes in processes and tasks. By augmenting human capabilities, automating processes, and enabling tasks that were once considered difficult or impossible, robotics is revolutionizing the industry.

Here are a few outstanding programs of robotics:

- Manufacturing and Industry: Robotics has been used in manufacturing since its inception. Robots are used to automate repetitive and precise tasks on assembly lines, improving efficiency, accuracy, and production rates. They perform a wide range of tasks, such as welding, painting, packaging, and assembly, in various industries from automotive to electronics.

- Healthcare: Robots can assist surgeons in delicate procedures with precision and consistency. They are used in minimally invasive surgeries, rehabilitation, and even in tasks like medication delivery, patient assistance, and monitoring. Robotic exoskeletons also help patients with mobility impairments.

- Logistics and Warehousing: Robotics optimizes warehouse processes by automating tasks such as sorting, picking, and packing goods. Autonomous mobile robots navigate warehouses, optimizing inventory management and order fulfillment. They can also perform material handling tasks in logistics.

- Agriculture: Robotics contributes to precision farming by employing drones and autonomous vehicles equipped with sensors to monitor crops, apply pesticides, and perform other agricultural tasks with precision, reducing waste, and increasing yields.

- Space Exploration: Robots are essential in space exploration, performing tasks in hazardous or inaccessible environments. Rovers explore celestial bodies, collect samples, and conduct experiments, expanding our understanding of outer space.

- Service and Hospitality: Service robots assist in various customer-facing roles, such as receptionists in hotels, assistants in retail environments, and even as companions for the elderly or people with disabilities, providing support and companionship.

- Education and Research: Robotics is used in educational settings to teach programming, engineering concepts, and as research platforms to explore AI, machine learning, and robotics itself.

They provide hands-on gaining knowledge of studies for college students of all ages.

- Defense and Security: In defense, robots are used for tasks such as bomb disposal, surveillance in dangerous areas, and reconnaissance. Unmanned aerial vehicles (UAVs) and unmanned ground vehicles (UGVs) aid in military operations while keeping human personnel out of harm's way.

Advancements in robotics, artificial intelligence, and sensor technology, these applications continue to evolve. Robotics remains at the forefront of technological innovation across various industries, offering a future with increased efficiency, safety, and capabilities.

Robots and Artificial Intelligence (AI) work together to enable robots to navigate, sense, and react to their surroundings using computer vision. The robots learn to perform tasks through machine learning which is a part of computer programming and AI. There are three types of AI used for robots, namely weak, strong, and specialized AI, depending on the tasks they need to perform. The collaboration between robots and AI has led to significant advancements in automation and robotics, enabling robots to perform a wide range of tasks with greater efficiency and accuracy.

CHAPTER 6

- AI Ethics and Regulations

AI ethics is a field that focuses on identifying and addressing the ethical issues that arise from the use of artificial intelligence. It seeks to optimize the beneficial impact of AI while reducing risks and adverse outcomes. Some of the ethical concerns that AI ethics addresses include data responsibility and privacy, fairness, explainability, transparency, inclusion, value alignment, accountability, trust, and technology misuse.

As companies increasingly rely on automation and data-driven decision-making, they are experiencing unforeseen consequences in some of their AI applications, particularly due to biased datasets and poor upfront research design.

This has led to the emergence of guidelines from the research and data science communities to address concerns around the ethics of AI. Leading companies in the field of AI have also taken an interest in shaping these guidelines to avoid reputational, regulatory, and legal exposure.

The ethical challenges of AI can be traced to the operational parameters of decision-making algorithms and AI systems. These operational characteristics can cause failures involving multiple human, organizational, and technological agents. Assigning responsibility and liability for the impact of AI behaviors becomes a difficult task due to the mix of human and technological actors involved.

This difficulty is captured in traceability as an overarching type of concern. As the appropriate expertise develops within the government industry, more AI protocols for companies to follow can be expected, enabling them to avoid any infringements on human rights and civil liberties.

As AI becomes increasingly prevalent throughout various industries, ethical concerns have arisen. To address these concerns, the academic community has established principles to guide ethical experimental research and algorithmic development. One such guide is the Belmont Report, which provides three main principles that serve as a framework for experiment and algorithm design.

The first principle is "Respect for Persons," which emphasizes the autonomy of individuals and the need for researchers to protect those with diminished autonomy. This could occur for a variety of reasons, such as illness, mental disability, or age restrictions. The principle focuses on the importance of obtaining informed consent from individuals, so they are aware of the potential risks and benefits of any experiment and can choose to participate or withdraw at any time.

The second principle is "Beneficence," which requires algorithms to "do no harm. This principle draws on healthcare ethics, where doctors take an oath to do no harm to their patients. In the context of AI, algorithms can amplify biases around race, gender, and political leanings, despite the intention to do good and improve a given system. Therefore, algorithms must be designed with the principle of beneficence in mind to ensure that they do not cause harm.

The third principle is "Justice," which deals with issues of fairness and equality. This principle raises questions about who should reap the benefits of experimentation and machine learning.

The Belmont Report offers five ways to distribute burdens and benefits: equal share, individual need, individual effort, societal contribution, and merit. By considering these factors, researchers can ensure that the distribution of benefits and burdens is as fair and equal as possible.

Governance is a critical process for organizations that use AI in their operations. It involves establishing internal policies and processes, hiring staff, and setting up systems to oversee the entire life-cycle of AI. The primary goal of governance is to ensure that AI systems operate in accordance with an organization's principles and values, meet stakeholders' expectations, and comply with relevant regulations.

A successful governance program should define the roles and responsibilities of people involved in the AI life-cycle. It should also educate them about building AI in a responsible way. Additionally, governance should establish processes for building, managing, monitoring, and communicating about AI and AI risks. Effective governance programs also leverage tools to improve AI's performance and trustworthiness throughout its life-cycle.

One of the most effective governance mechanisms is an AI Ethics Board. IBM's AI Ethics Board, for instance, comprises diverse leaders from across the business. It gives a centralized governance, review, and decision-making technique for IBM ethics rules and practices.

Organizations can also use principles and focus areas to guide their approach to AI ethics. These principles can be applied to products, policies, processes, and practices throughout the organization to help enable trustworthy AI. They should be supported by focus areas, such as explainability or fairness, around which standards can be developed, and practices can be aligned.

Building AI with ethics at its core can have tremendous potential to impact society positively. In the healthcare sector, for example, AI has already been integrated into radiology. However, it is crucial to assess and mitigate the potential risks associated with AI's uses, beginning in the design phase. The conversation around AI ethics is therefore essential to ensure that AI is used responsibly and ethically.

Organizations that promote AI Ethics
As the use of artificial intelligence (AI) continues to expand, concerns about the ethical implications of its implementation

have also grown. In response to this, several organizations have emerged to promote ethical conduct in the field of AI.

One such organization is AlgorithmWatch, a non-profit that focuses on promoting an explainable and traceable algorithm and decision process in AI programs. Their website provides a wealth of resources for those seeking more information on AI ethics.

Another non-profit, the AI Now Institute at New York University, conducts research on the social implications of AI. Their website offers a variety of publications and resources related to AI ethics and policy.The Defense Advanced Research Projects Agency (DARPA), a research and development agency of the US Department of Defense, is also committed to promoting explainable AI and AI research.

The Center for Human-Compatible Artificial Intelligence (CHAI) is a collaboration of various institutes and universities that seeks to promote trustworthy AI and provable beneficial systems. Their website provides information on their research and initiatives related to AI ethics.

Finally, the National Security Commission on Artificial Intelligence (NASCAI) is an independent commission focused on advancing the development of AI to address the national security and defense needs of the United States. Their website provides information on their reports and recommendations related to AI and national security.

By promoting ethical AI practices, these organizations are helping to ensure that AI is developed and used in a responsible and beneficial way for all.

IBM Insight on AI Ethics

IBM is a leading technology company that has established its perspective on AI ethics. It has developed a set of principles called the "Principles of Trust and Transparency" to provide guidance to its clients on the values that underpin its AI development. IBM's approach is centered on three core principles: augmenting human intelligence, ownership of data and insights, and transparency and explainability of AI systems.

IBM believes that AI should be used to support human intelligence and not replace it. The company is committed to investing in global initiatives to promote skill training around AI technology to support the transition of workers. IBM's clients are assured that they own their data, and the company will not provide government access to client data for any surveillance programs. IBM is also committed to protecting the privacy of its clients.

In addition, IBM has developed five pillars to guide the responsible adoption of AI technologies. These include explainability, fairness, robustness, transparency, and privacy. IBM believes that AI systems should be transparent, explainable, and fair. The company also recognizes the importance of safeguarding consumers' privacy and data rights.

Overall, IBM's approach to AI ethics is centered on transparency, accountability, and responsibility. The company believes that it is essential to be clear about the data used to train AI systems and how their algorithms make recommendations.

Additionally, AI systems must be actively defended from adversarial attacks to minimize security risks and promote confidence in the system outcomes.

The emergence of artificial intelligence (AI) has brought forth various implications, including the need for a regulatory framework to address its impact on society. Several approaches have been proposed to address this issue, but the question remains: should a new government agency be established to regulate AI, or should existing agencies be strengthened instead?

Microsoft and other companies have proposed the creation of a new government AI agency to address the challenges posed by highly capable AI foundation models. The agency could bring about resources and expertise. However, history has shown that a new agency can lead to "gaming" of regulation, regulatory "capture" by incumbents, agency "mission creep," and slow, bureaucratic decision-making.

IBM and Google, on the other hand, propose that existing agencies' competencies be strengthened to address AI while leveraging their domain expertise. This approach could be supplemented with a multi-stakeholder approach, which has been used for Internet governance. Expert advisory committees are already employed in trade and cybersecurity.

A new White House coordinating office could also help bring focus, consistency across agencies, and high-level political leadership. On a technical level, agencies like NIST can provide a common technical approach to defining and measuring key concepts like data quality, bias, explainability, and auditability.

Private businesses are better positioned to address certain tasks, such as safe model design, testing, publishing model capabilities, and internal AI ethics boards. Other market-based approaches include third-party auditors to test AI or insurance-based approaches.

Regulation regularly works satisfactory while it makes use of carrots (incentives, secure harbors) in addition to sticks (penalties). Google and others recognition on "right AI" consequences which include an AI-geared up workforce, making an investment in AI innovation and competitiveness, and assisting huge AI adoption. The EU AI Act takes a strong enforcement approach and has big financial penalties. U.S. AI efforts were by and large voluntary, including the White House July 21, 2023 commitments from AI companies.

One of the significant developments in the scope of AI regulation is the consideration of not only the model developer, but also the applications and IT infrastructure it runs on. This expansion of regulation is necessary to ensure the safe and responsible use of AI in high-risk areas such as critical infrastructure. The licensing requirements for AI data centers, as proposed by some, are a positive step towards achieving this goal.

The international harmonization of AI rules is being pursued through various initiatives and organizations, such as the EU AI Act, the G7's "Hiroshima AI process," the OECD's AI principles, and the United Nations' proposed new UN AI advisory body. While conflicting national rules are likely to persist, these initiatives offer hope for greater consistency and cooperation in the future. International technical standards developed via ISO and IEEE also show promise in this regard.

The likely outcome for AI regulation in the United States is a decentralized, bottom-up approach that may have gaps and inconsistencies. However, there are several positive developments underway. For example, the administration's AI "bill of rights" spans different sectors and provides voluntary commitments and executive orders to address sensitive areas such as healthcare, financial services, workforce practices, and child safety.

While a broad-based national AI law like the EU Act is unlikely in the near future, executive branch agencies are likely to move forward with existing authorities in the absence of new laws.

Private tech companies are also advancing their own responsible AI initiatives to avoid tougher government action. These initiatives are largely voluntary and may be insufficient, but they demonstrate a growing sense of responsibility and commitment to ethical standards beyond legal requirements. Federal spending on AI and AI research is expanding, with funds coming with market-shaping rules, as the US uses its buying power in crucial areas such as health, education, national security, and public safety. The National Science Foundation is funding 25 AI research institutes in part to boost national economic competitiveness.

Executive orders are likely to be issued to limit AI bias and risks in federal agency programs. Additional orders may also focus on enhancing adoption of AI in federal IT to improve citizen services and strengthen AI security, with uneven success. While the decentralized structure of the US government and political differences may slow down the pace of AI regulation, there is also a possibility of a big AI-related failure sparking strong government action. Overall, the growing awareness and focus on AI regulation at all levels is a positive sign, and the various measures being taken demonstrate a commitment to ensuring that AI Is evolved and deployed in a manner that blessings society as a whole.

As AI technology continues to advance and become more widely integrated into our daily lives, it is crucial that we examine and address its ethical and societal implications. The papers published in "AI for People" represent important contributions to this ongoing conversation, but there is still much work to be done in order to fully understand the complex issues that arise from the development and deployment of AI.

One topic for future discussion is the expansion of AI to encompass not only people, but also the world and the universe as a whole. Additionally, the impact of AI on the workplace and its potential for human adaptation, as well as ethical considerations for workplace AI, are important topics to explore.

Trustworthy AI is another area of focus, as it is critical to ensure that autonomous systems are programmed in a way that is transparent, accountable, and trustworthy. The role of explainability in supporting trust is also a crucial aspect to consider.
The use of AI for social good is another important topic, including aligning

academic journal ratings with the United Nations Sustainable Development Goals (SDGs) and investing in AI for social good. Examining the legal implications of AI, including the legal status of intelligent service robots and the use of AI in criminal justice, is also an area of interest. Exploring the intersection of AI and culture, including cultural robotics and the cultural history of consciousness, is another important topic.

Investigating biases in AI, such as implicit biases and stereotypes, age discrimination and exclusion, and bias and discrimination in job advertisements, is a crucial area of focus as well.

Finally, addressing various technological issues related to AI, such as developing a practical ethical methodology for integrating AI into the industry, algorithmic fairness, and the importance of transparency in AI operations, is essential.

It is critical that we approach the development and deployment of AI with a clear understanding of its limitations and ethical implications. By building on the work done so far and exploring these important topics, we can work towards a future in which AI is developed and deployed in a way that benefits all members of society.

It is important for AI engineers and developers to embrace and practice a responsible approach to the development of AI technologies, taking into account not only the technical aspects of Their structures, but additionally the moral implications in their use. Additionally, developers should consider various delegated responsibilities, such as data collection, privacy, and intellectual property.

Addressing the ethical implications of AI should not be limited to individual technology development, but should also consider the broader context of government policies and civil society initiatives. As such, a common framework of principles and standards is needed to ensure accountability, fairness, and transparency in the use of AI. Governments should create policies and guidelines to ensure that AI is used responsibly while enabling innovation and progress.

Hackers launch millions of attacks each year, ranging from simple malware assaults to sophisticated network breaches. These threats can have a profound impact on businesses, causing damage to their networks and stealing sensitive data.

Unfortunately, many businesses lack the expertise to recognize and prevent these threats before they occur.

- Flaw Identification: One way AI can assist businesses is by detecting flaws in their system, such as buffer overflows. These occur when programs consume more data than usual, causing data to spill over into other parts of the system. These errors can also be caused by programming language flaws and can be difficult for humans to detect. However, AI can identify these issues in real-time and prevent future threats.

- Threat Prevention: AI can also be used to prevent cyber threats before they occur. Cybersecurity vendors are constantly developing AI technology that can detect flaws in the system and install additional firewalls and rectify code faults that lead to dangers. If anyone attempts to exploit these issues, AI will instantly exclude them from the system.

- Responding to Threats: In the event that a threat does enter the system, AI can detect unusual behavior and create an outline of viruses or malware. AI can then take appropriate action against the threat, such as removing the infection, repairing the fault, and administering the harm done.

- Recognizing Uncharacterized Action: AI can detect unusual behavior in a system by continually scanning and gathering data. AI identifies illegal access and employs particular elements to determine whether it represents a genuine threat or a fabricated warning. Machine learning is also used to help AI determine what is and is not aberrant behavior. AI can point out anything wrong with the system, allowing businesses to take appropriate action.

In addition to cybersecurity, AI is also being used in the automotive sector to improve production, supply chain optimization, passenger and driver experience, inspections, and quality control. For example, AI can be used to boost productivity in automobile assembly, design cars, and improve performance using sensors. AI can also assist with supply chain forecasting, routing difficulties, volume forecasts, and other concerns. Finally, AI can enhance the customer experience by decreasing distractions for drivers, analyzing driving behaviors, and providing customized accessibility for passengers.

In summary, artificial intelligence is becoming an increasingly important tool for businesses looking to prevent cyber threats and improve their operations. With its ability to detect flaws, prevent threats, respond to incidents, and recognize unusual behavior, AI is quickly becoming an essential component of modern cybersecurity and automotive production.Artificial Intelligence (AI) is becoming increasingly important in the realm of data security, which is critical for any tech-oriented company that deals with confidential data. This includes customer data such as credit card information, as well as organizational secrets kept online.

As the world becomes more connected, cyberattacks are becoming more sophisticated, and security teams will need to rely on AI solutions to keep their systems and data secure.

One of the most important applications of AI in data security is the detection of unknown threats. AI systems can analyze vast amounts of data and recognize patterns that may indicate a security breach. By using machine learning algorithms, AI can identify and respond to potential threats before they become a serious problem. This is particularly useful in situations where traditional security methods may not be enough to detect new and evolving threats.

AI can also be used to improve the accuracy of threat detection. By analyzing data from multiple sources, AI systems can provide a more complete picture of potential threats and help security teams take action more quickly. Additionally, AI can automate routine security tasks, such as monitoring network traffic and log files, freeing up security analysts to focus on more complex threats.

Overall, AI is critical for any organization that wants to keep their data and systems secure. As cyberattacks become more sophisticated, businesses need to be proactive and use the latest tools and technologies to protect their assets. With AI-powered security solutions, companies can stay one step ahead of potential threats and keep their data safe from harm.

CHAPTER 7

- Generative AI and AI Art

Generative AI is an exciting technology that is redefining what is possible. With the ability to analyze vast datasets and generate new ones, this technology can save programmers hours of time and help marketers create automations without writing a line of code. While the concept of generative AI is not new, recent advances in neural network models and the availability of large amounts of data have led to a significant qualitative leap in its capabilities.

Projects like Google's DeepDream and GANs paved the way for generative AI, but it is models like OpenAI's GPT-3 that have taken it to the next level, with the ability to create coherent text, programming code, poems, and even dialogue for video games.

As software complexities continue to grow, the convergence of generative AI and software architecture holds the potential to transform how we design and develop software.

From automating mundane tasks to creating resilient software systems, the possibilities are tantalizing. However, there are also challenges to be addressed in this AI-augmented future, and this paper aims to shed light on both the opportunities and the pitfalls, ensuring that domain experts and readers are well-equipped for what lies ahead.

The integration of AI with software architecture has led to a significant transformation in the field. With the transition from rule-based systems to deep learning models, AI has introduced generative models such as GANs, VAEs, and Transformer models that can create content with unprecedented accuracy. These models are now at the forefront of the next wave of software architecture, promising a future where designs are not just created but evolved

Initial Programming Designs for Generative AI and Problem-Solving Techniques for Generative AI

Generative AI has become a game-changer in software design, where it can generate initial software blueprints based on high-level requirements, predict and rectify potential vulnerabilities, and create adaptable and resilient software systems. These models can revolutionize software design by enhancing modularity, reusability, and adaptability, and improving code quality.

Generative AI has also influenced various methodologies in software architecture, such as modeling, designing, trade-offs, decisions, and patterns and principles. It can predict and generate optimal models, suggest design improvements, weigh in on architectural decisions, and suggest or even generate new patterns for recurrent problems. Traditional practices are now being re-evaluated in the light of generative AI techniques, promising a future where software architecture co-evolves with AI.

Deep learning models like GANs have found applications in software design, where they can learn from existing architectural patterns to generate new ones. Techniques for automated code generation and architectural optimization are also being developed, which use AI algorithms to generate code snippets or entire modules based on specified requirements.

AI-powered tools like ChatGPT are steering the transformation of software development, redefining how software solutions are conceived, designed, and realized. Generative AI can analyze both functional and non-functional requirements to suggest the most suitable design patterns, recommend optimal technology stacks, transform textual descriptions into architectural diagrams, and provide a detailed analysis of trade-offs based on historical data, case studies, and best practices.

Google Cloud offers a range of generative AI services, including Vertex AI, which allows users to interact with and embed foundation models into their applications without requiring any machine learning expertise. Developers can access foundation models on Model Garden, tune models via Generative AI Studio, or use models within a data science notebook. Additionally, Google Cloud's Vertex AI Search and Conversation service provides developers with the ability to build generative AI-powered search engines and chatbots quickly.

Google's experts can also assist organizations looking to implement generative AI solutions. Consulting services are available to help with creating new content, discovering trends and gaining insights from data, summarizing for faster decision-making, and automating solutions and processes.

Applications of Generative AI

Generative AI is a groundbreaking technology that has found diverse applications in various industries. It has the ability to create new content and simulate human-like creativity, making it a valuable tool for businesses and researchers alike. Let's look at some of the notable applications of Generative AI in more detail.

1. Art and Creativity: Generative AI is extensively used in art and creative fields. It generates visual art, paintings, and images that are sometimes indistinguishable from human-created art. It also aids in music composition, generating new melodies, and even writing poetry or literature. This technology is revolutionizing the way artists and creatives approach their work and helping them to produce new, innovative content.

2. Content Creation and Media: Generative AI generates content for video games, special effects in movies, and

virtual environments. It's also used to create realistic voice synthesis, deep fakes, and personalized content generation. This technology is helping media and entertainment companies to create more engaging and immersive experiences for their audiences.

3. Fashion and Design: AI assists in fashion design by generating new clothing designs, predicting trends, and personalizing fashion recommendations for customers based on their preferences. This technology is transforming the fashion industry by helping designers to create new fashion lines and provide personalized recommendations that meet the needs of their customers.

4. Healthcare and Biology: Generative models contribute to drug discovery by predicting molecular structures and interactions. They also generate synthetic data for medical imaging and assist in generating synthetic biological data for research purposes. This technology is

helping researchers to develop new drugs and treatments faster and more efficiently.

5. Simulation and Training: In robotics and autonomous systems, Generative AI helps simulate environments for training purposes. It generates synthetic data for training machine learning models, particularly in scenarios where real data is limited or expensive to acquire. This technology is helping researchers to develop more advanced and capable robots and autonomous systems.

6. Natural Language Processing (NLP) and Writing: AI generates human-like text, assists in language translation, and helps in content summarization. It's employed in chatbots, dialogue generation, and content creation for various purposes. This technology is revolutionizing the way we communicate and helping businesses to provide better customer service and engagement.

7. Finance and Business: Generative AI aids in generating financial models, forecasting, and fraud detection. It's used in algorithmic trading, risk assessment, and generating synthetic financial data for testing systems. This technology is helping businesses to make better financial decisions and detect fraud more accurately.

8. Personalization and Recommendation Systems: AI generates personalized recommendations in e-commerce, content streaming platforms, and social media. It predicts user preferences, leading to more tailored and engaging experiences. This technology is helping businesses to provide personalized recommendations that meet the needs and preferences of their customers.

9. Generative Design and Architecture: In architecture and design, AI generates innovative design concepts, floor plans, and assists in urban planning by simulating cityscapes or architectural layouts. This technology is helping architects and designers to create more

innovative and sustainable designs that meet the needs of communities.

10. Scientific Research and Exploration: Generative AI contributes to scientific research by generating hypotheses, analyzing complex datasets, and simulating scientific experiments in fields like physics, astronomy, and climate modeling. This technology is helping researchers to make breakthroughs in scientific research and exploration.

These applications showcase the versatility of Generative AI, transforming industries, fostering innovation, and pushing the boundaries of what's possible in creativity, problem-solving, and data synthesis. As the field continues to advance, it's expected to find even more diverse and impactful applications across various domains.

These applications showcase the versatility of Generative AI, transforming industries, fostering innovation, and pushing the boundaries of what's possible in creativity, problem-solving, and data synthesis. As the field continues to advance, it's expected to find even more diverse and impactful applications across various domains.Generative AI is a groundbreaking technology that has found diverse applications in various industries.

It has the ability to create new content and simulate human-like creativity, making it a valuable tool for businesses and researchers alike. Let's look at some of the notable applications of Generative AI in more detail.

Art and Creativity: Generative AI is extensively used in art and creative fields. It generates visual art, paintings, and images that are sometimes indistinguishable from human-created art. It also aids in music composition, generating new melodies, and even writing poetry or literature. This technology is revolutionizing the way artists and creatives approach their work and helping them to produce new, innovative content.

Fashion and Design: AI assists in fashion design by generating new clothing designs, predicting trends, and personalizing fashion recommendations for customers based on their preferences. This technology is transforming the fashion industry by helping designers to create new fashion lines and provide personalized recommendations that meet the needs of their customers.

Healthcare and Biology: Generative models contribute to drug discovery by predicting molecular structures and interactions. They also generate synthetic data for medical imaging and assist in generating synthetic biological data for research purposes. This technology is helping researchers to develop new drugs and treatments faster and more efficiently.

Simulation and Training: In robotics and autonomous systems, Generative AI helps simulate environments for training purposes. It generates synthetic data for training machine learning models, particularly in scenarios where real data is limited or expensive to acquire. This technology is helping researchers to develop more advanced and capable robots and autonomous systems.

Natural Language Processing (NLP) and Writing: AI generates human-like text, assists in language translation, and helps in content summarization. It's employed in chatbots, dialogue generation, and content creation for various purposes. This technology is revolutionizing the way we communicate and helping businesses to provide better customer service and engagement.

Finance and Business: Generative AI aids in generating financial models, forecasting, and fraud detection. It's used in algorithmic trading, risk assessment, and generating synthetic financial data for testing systems. This technology is helping businesses to make better financial decisions and detect fraud more accurately.

Personalization and Recommendation Systems: AI generates personalized recommendations in e-commerce, content streaming platforms, and social media. It predicts user preferences, leading to more tailored and engaging experiences. This technology is helping businesses to provide personalized recommendations that meet the needs and preferences of their customers.

Generative Design and Architecture: In architecture and design, AI generates innovative design concepts, floor plans, and assists in urban planning by simulating cityscapes or architectural layouts. This technology is helping architects and designers to create more innovative and sustainable designs that meet the needs of communities.

Scientific Research and Exploration: Generative AI contributes to scientific research by generating hypotheses, analyzing complex datasets, and simulating scientific experiments in fields like physics, astronomy, and climate modeling. This technology is helping researchers to make breakthroughs in scientific research and exploration.

These applications showcase the versatility of Generative AI, transforming industries, fostering innovation, and pushing the boundaries of what's possible in creativity, problem-solving, and data synthesis. As the field continues to advance, it's expected to find even more diverse and impactful applications across various domains.

Challenges and Limitations of Generative AI

Generative AI is an incredibly powerful and versatile technology, but it still faces several challenges and limitations that impact its development and applications. One of the key challenges is generating high-quality and realistic outputs, which is difficult to achieve consistently across different domains such as images, text, or music. Another challenge is avoiding dataset bias and ensuring that models trained on diverse and representative data can generalize well to unseen data.

One of the biggest limitations of Generative AI is the lack of interpretability and explainability, making it difficult to understand how and why a model generates specific outputs.

Additionally, some models suffer from mode collapse and lack diversity in their outputs, particularly in creative domains like art or music. Training and deploying Generative AI models often require significant computational resources, including high-performance hardware and massive amounts of data, which can be a challenge.

Another critical concern is the ethical and legal considerations surrounding the generation of realistic content. Addressing ethical guidelines and legal regulations for responsible use is crucial, particularly regarding deepfakes and synthetic media's potential misuse for misinformation or fraud. Ensuring robustness and security against adversarial attacks is also essential, especially in safety-critical applications.

Some models struggle with understanding and maintaining long-term dependencies or context in generated sequences, impacting the coherence and relevance of generated outputs, particularly in language generation or storytelling. Furthermore, Generative AI models can amplify existing biases present in the training data, which is a significant concern in decision-making applications.

Developing appropriate regulatory frameworks and ethical guidelines for governing Generative AI's use and ensuring responsible deployment pose challenges due to the rapid evolution of the field. Addressing these challenges involves interdisciplinary efforts, including advancements in machine learning algorithms, data curation practices, interpretability techniques, ethical considerations, and regulatory frameworks. Overcoming these limitations will contribute to the responsible and beneficial integration of Generative AI into various domains.

Overview of AI Art

AI art is an emerging field that combines artificial intelligence and artistic expression. It involves the use of algorithms and machine learning models to generate visual or auditory content that is indistinguishable from human-created art. This type of art explores the creative process by leveraging AI's ability to learn patterns from large datasets and produce novel outputs.

AI art relies on various machine learning techniques such as Generative Adversarial Networks (GANs), Variational Autoencoders (VAEs), recurrent neural networks, and deep learning architectures to generate artistic content. These models learn patterns from existing artworks and datasets to create new, original pieces.

Style transfer techniques use AI to apply the artistic style of one image onto another, blending them to create unique compositions. Neural style transfer and similar methods enable the transformation of photographs or images into various artistic styles.

AI art often involves collaboration between artists and AI systems, where artists leverage AI tools to augment their creative process. AI acts as a co-creator, assisting artists in exploring new techniques, generating ideas, or aiding in the creation of complex pieces.

AI facilitates the creation of interactive and digital art installations that respond to the viewer's input or environmental cues. These installations use machine learning algorithms to create dynamic, responsive, and immersive artistic experiences.

AI art raises questions about authorship, creativity, and the definition of art. Debates around whether AI-generated art can possess true creativity or if it merely mimics existing styles continue.

Additionally, ethical concerns, including ownership of AI-generated artworks and the potential for misuse or manipulation, are areas of ongoing discussion.

AI-generated artworks have gained recognition in the art world, with some pieces being exhibited in galleries, sold at auctions, and integrated into various cultural spheres. This type of art challenges notions of artistic authorship, creativity, and the relationship between technology and art.

Overall, AI art represents an evolving landscape where technology merges with creativity, opening up new possibilities for artistic expression, collaboration, and contemplation about the nature of art in the digital age.

Techniques and Tools for AI Art Creation

AI-powered art creation involves a wide range of techniques and tools that leverage machine learning algorithms and generative models. These methods enable artists and developers to create unique, original content across various mediums, including images, music, videos, and text.

One of the most widely used techniques in AI art creation is Generative Adversarial Networks (GANs). GANs consist of two neural networks, a generator and a discriminator, that are trained in an adversarial manner. By learning from a dataset, GANs generate new content based on the learned patterns, producing realistic images.

Another technique is Variational Autoencoders (VAEs), which enable the generation of new, original content by learning latent representations of input data. VAEs encode input data into a lower-dimensional latent space, allowing for controlled manipulation and generation of new outputs.

Style transfer techniques, such as neural style transfer, enable the application of artistic styles from one image onto another. These methods use convolutional neural networks to blend the style of one image with the content of another, creating novel artistic compositions.

Recurrent Neural Networks (RNNs) and Natural Language Processing (NLP) techniques are utilized for generating textual content, including poetry, stories, and dialogue. These models learn from large text corpora and generate coherent and contextually relevant text.

DeepDream is a visualization technique used to create dream-like visuals by enhancing patterns within images. Neural art tools, like DeepArt or RunwayML, allow users to create artistic images using pre-trained models and custom inputs.

Interactive art platforms, such as Google's Magenta Studio or RunwayML, provide interactive environments for artists to experiment with AI-powered tools. These platforms offer user-friendly interfaces for creating music, art, or videos using AI algorithms.

Creative coding libraries, such as TensorFlow, PyTorch, and Keras, provide a foundation for developing AI art projects. Artists and developers can access pre-trained models, build custom models, and experiment with various machine learning techniques.

Curating diverse and representative training datasets is crucial for training AI models in art creation. Data augmentation techniques are used to enhance datasets by manipulating existing images or content to create variations.

Collaborative projects between artists and AI systems involve using AI tools to assist or co-create art. This approach combines human creativity with AI's generative capabilities to explore new artistic directions.

Finally, ethical considerations and guidelines are crucial for artists and developers working with AI in art creation. They need to consider ethical implications, such as authorship, ownership, and the responsible use of AI-generated content.

The advancements in artificial intelligence have brought forth several AI-powered art generation systems such as Midjourney, DALL-E, Google Deep Dream, and DeepArt.io. These systems employ machine learning and generative art techniques to produce creative art based on user input. Midjourney and DALL-E generate more artistic and photorealistic art, respectively, while Google Deep Dream applies algorithmic filters to make images appear more surreal. Similarly, DeepArt.io provides users with several styles to apply to their original uploaded images and also allows customization options.

These AI-generated artworks have already made their way into art exhibitions and competitions, sparking controversy and raising questions about the definition of art.

Examples of AI Art

I-generated art can be found across a diverse range of mediums, including visual art, music, literature, and interactive installations. Some notable examples include:

- AI-generated visual art, such as the portrait "Edmond de Belamy" created by Obvious or Microsoft's "The Next Rembrandt", which mimics the style and techniques of the famous artist.

- AI-generated music compositions, such as Google's Magenta project that produces compositions ranging from classical to experimental genres, and AIVA, an AI composer that creates symphonies in collaboration with human musicians.

- AI-generated text and poetry, such as Botnik Studios' predictive text algorithms that generate poetry, stories, and scripts, and OpenAI's GPT-3 language model that creates coherent and contextually relevant content.

- Interactive and digital art installations, such as Mario Klingemann's neural art and Refik Anadol's data sculptures that use AI algorithms to create immersive visual experiences.

- AI-powered image manipulation and style transfer, such as DeepArt's online platform that applies artistic styles to user-submitted images, and RunwayML's tools for real-time image manipulation and creative applications.

- AI-generated literature, such as AI Dungeon's interactive stories and adventures that respond dynamically to users' input, and Google Arts & Culture PoemPortraits, which generates poems based on user-submitted words.

These examples demonstrate the extent to which AI has emerged as an important part of the creative process, working alongside human creators to produce innovative and unique works of art.
Future Developments in Generative AI and AI Art

The future of Generative AI and AI art is expected to bring about exciting possibilities, driven by ongoing advancements in technology and creative exploration. Here are some anticipated developments:

One of the expected developments is an increase in the realism and fidelity of AI-generated content across various domains. This is expected to blur the lines between AI-generated content and human-created art. Another expected development is the improvement of AI's ability to produce truly original and creative content. The focus will be on going beyond mimicking existing styles and exploring novel artistic expressions.

Collaboration between artists, technologists, and AI systems is also expected to deepen, resulting in new artistic forms, interactive installations, and immersive experiences that integrate AI and human creativity seamlessly. As AI-generated content becomes more prevalent, there will likely be an emphasis on developing robust ethical guidelines, copyright laws, and regulations governing ownership and responsible use of AI-generated artworks.

AI art is expected to evolve towards personalized experiences, allowing individuals to interact with and influence artworks, creating dynamic and tailored artistic encounters. AI tools will continue to augment human creativity, offering artists new tools and techniques for ideation, experimentation, and expanding artistic boundaries.

Further developments in neural network architectures, attention mechanisms, and memory-augmented models will enhance AI's ability to understand context, produce long-form content, and maintain coherence in generated outputs. AI will likely play a role in preserving cultural heritage by reconstructing or restoring artworks, texts, and artifacts. Additionally, AI-based educational tools will continue to evolve, fostering creativity and learning in new ways.

User-friendly AI art tools and platforms will become more accessible to a broader audience of artists and creators, enabling them to explore AI-driven creativity without extensive technical expertise.

Finally, AI's creative potential will extend to new artistic mediums and interdisciplinary domains, pushing the boundaries of what defines art and how it's created.

These developments in Generative AI and AI art promise a future where artificial intelligence not only assists in creative endeavors but also contributes to shaping new artistic landscapes, fostering innovation, And redefining the connection among era and creative expression.

Exploring the Dark Side of AI Art: The Intersection of Creativity and Machine Intelligence

The emergence of AI art has raised several ethical concerns and potential negative implications that need to be addressed.

These issues are complex and multifaceted, touching on different aspects of AI-generated content.

One of the primary concerns is the potential for AI-generated content to be used for manipulation and misinformation. Deepfakes, forged documents, or misleading media created using AI can undermine trust and truthfulness and have a significant impact on society.

Determining the ownership and authorship of AI-generated art is another challenge. The question of who holds the rights and credits for the content blurs the lines between human creativity and machine-generated outputs.

AI models trained on biased datasets can perpetuate existing biases and stereotypes present in the data. This can lead to the reinforcement of societal biases in AI-generated content, affecting representation and fairness.

AI-powered tools for facial recognition or surveillance can infringe on privacy rights, raising concerns about the use of AI-generated content for tracking or identifying individuals without consent.

AI-driven personalized content creation might aim to manipulate emotions or behaviors, potentially impacting mental health or influencing decision-making without individuals' awareness. There are also concerns that algorithms might control creativity, shaping artistic trends and restricting the diversity of creative expression. This could lead to homogenized or algorithmically guided art forms, limiting the scope of creativity.

The unforeseen consequences of AI-generated content in cultural contexts, legal disputes, or societal norms also present challenges that need to be addressed.

Finally, over-reliance on AI for creative processes might diminish human creativity, originality, and the diversity of artistic expressions, potentially impacting the artistic landscape.

Addressing these challenges requires a multidisciplinary approach, encompassing ethical guidelines, regulatory frameworks, transparency in AI systems, and raising awareness about the implications of AI-generated art. Striking a balance between fostering innovation and addressing the ethical, social, and cultural implications of AI art is crucial for its responsible and beneficial integration into society.

Conclusion

We stand at the precipice of a future illuminated by the transformative power of AI. As we explore the untapped potential that AI holds, we are empowered to transcend boundaries, imagine new worlds, and innovate beyond the confines of our imagination. The promise of AI extends far beyond mere automation or optimization, igniting the fires of creativity and innovation within us. It is not the harbinger of an age devoid of human touch, but a companion in our pursuit of a more vibrant, more empathetic, and more innovative world.

As we navigate the ethical landscapes that AI unveils, we must remain steadfast in our ethical compass and make decisions that will positively impact society for generations to come. The intersection of AI and humanity is not a collision but a fusion, a symbiosis that enriches both parties. As AI augments our capabilities, it elevates our potential, freeing us from the mundane and empowering us to focus on endeavors that resonate with the essence of our humanity - creativity, empathy, and exploration.

AI is not the end but a gateway to an era where humanity, empowered by its creations, crafts a future beyond imagination.

With the limitless horizon of AI at our disposal, we can create a future that embodies the best of human potential, augmented by the power of artificial intelligence. Let us embrace the transformative power of AI and carry forth the lessons learned, the ethical principles upheld, and the imagination ignited, as we shape the future of humanity.

www.ingramcontent.com/pod-product-compliance
Lightning Source LLC
Chambersburg PA
CBHW071203290526
45796CB00008B/125